HEALTHINFO 2023

The Eighth International Conference on Informatics and Assistive Technologies for Health-Care, Medical Support and Wellbeing

November 13th – 17th, 2023

Valencia, Spain

HEALTHINFO 2023 Editors

Jaime Lloret Mauri, Universitat Politecnica de Valencia, Spain

Printed from e-media with permission by:

Curran Associates, Inc.
57 Morehouse Lane
Red Hook, NY 12571

CURRAN ASSOCIATES INC.
proceedings
.com

Some format issues inherent in the e-media version may also appear in this print version.

Printed with permission by Curran Associates, Inc. (2024)

International Academy, Research, and Industry Association (IARIA)
412 Derby Way
Wilmington, DE 19810

Phone: (408) 893-6407
Fax: (408) 527-6351

petre@iaria.org

Additional copies of this publication are available from:

Curran Associates, Inc.
57 Morehouse Lane
Red Hook, NY 12571 USA
Phone: 845-758-0400
Fax: 845-758-2633
Email: curran@proceedings.com
Web: www.proceedings.com

Table of Contents

HEALTHINFO 2023

Forward

The Eighth International Conference on Informatics and Assistive Technologies for Health-Care, Medical Support and Wellbeing (HEALTHINFO 2023), held on November 13 - 17, 2023 in Valencia, Spain, tackles with particular aspects belonging to health informatics systems, health information, health informatics data, health informatics technologies, clinical practice and training, and wellbeing informatics in terms of existing and needed solutions.

The progress in society and technology regarding the application of systems approaches information and data processing principles, modeling and information technology, computation and communications solutions led to a substantial improvement of problems in assistive healthcare, public health, and the everyday wellbeing. While achievements are tangible, open issues related to global acceptance, costs models, personalized services, record privacy, and real-time medical actions for citizens' wellbeing are still under scrutiny.

We take here the opportunity to warmly thank all the members of the HEALTHINFO 2023 technical program committee as well as the numerous reviewers. The creation of such a broad and high quality conference program would not have been possible without their involvement. We also kindly thank all the authors that dedicated much of their time and efforts to contribute to the HEALTHINFO 2023. We truly believe that thanks to all these efforts, the final conference program consists of top quality contributions.

This event could also not have been a reality without the support of many individuals, organizations and sponsors. We also gratefully thank the members of the HEALTHINFO 2023 organizing committee for their help in handling the logistics and for their work that is making this professional meeting a success.

We hope the HEALTHINFO 2023 was a successful international forum for the exchange of ideas and results between academia and industry and to promote further progress in health informatics research. We also hope that Valencia provided a pleasant environment during the conference and everyone saved some time for exploring this beautiful city

HEALTHINFO 2023 General Chair

Jaime Lloret Mauri, Polytechnic University of Valencia, Spain

HEALTHINFO 2023 Steering Committee

Shada Alsalamah, King Saud University, Saudi Arabia
Nelson P. Rocha, University of Aveiro, Portugal

HEALTHINFO 2023 Publicity Chair

Lorena Parra Boronat, Universitat Politecnica de Valencia, Spain
Sandra Viciano Tudela, Universitat Politecnica de Valencia, Spain
Jose Miguel Jimenez, Universitat Politecnica de Valencia, Spain

HEALTHINFO 2023

Committee

HEALTHINFO 2023 General Chair

Jaime Lloret Mauri, Polytechnic University of Valencia, Spain

HEALTHINFO 2023 Steering Committee

Shada Alsalamah, King Saud University, Saudi Arabia
Nelson P. Rocha, University of Aveiro, Portugal

HEALTHINFO 2023 Publicity Chair

Lorena Parra Boronat, Universitat Politecnica de Valencia, Spain
Sandra Viciano Tudela, Universitat Politecnica de Valencia, Spain
Jose Miguel Jimenez, Universitat Politecnica de Valencia, Spain

HEALTHINFO 2023 Technical Program Committee

Djafar Ould Abdeslam, University of Haute Alsace, France
Sherif Abdelwahed, Virginia Commonwealth University, USA
Somayeh Abedian, Ministry of Health and Medical Education, Tehran, Iran
Miriam Allalouf, Azrieli College of Engineering Jerusalem - JCE, Israel
Jens Allmer, Hochschule Ruhr West, University of Applied Sciences, Germany
João R. Almeida, University of Aveiro, Portugal / University of A Coruña, Spain
Shada Alsalamah, King Saud University, Saudi Arabia
Mustapha Aouache, Telecom Division - Centre de Développement des Technologies Avancées (CDTA), Algiers, Algeria
Khalfalla Awedat, Pacific LutheranUniversity, USA
Mana Azarm, University of Ottawa, Canada
Nabil Georges Badr, Higher Institute of Public Health – USJ, Beirut, Lebanon
Panagiotis D. Bamidis, Aristotle University of Thessaloniki, Greece
Hugo Barbosa, Lusofona University of Porto / Faculty of Engineering of the University of Porto, Portugal
Fabio Baselice, University of Naples Parthenope, Italy
Arriel Benis, Holon Institute of Technology, Israel
Ahmed Bentajer, National School of Applied Sciences | Abdelmalek Essaad University, Tetouan, Morocco
Vilmos Bilicki, University of Szeged, Hungary
Amine Boufaied, ISITCom | University of Sousse, Tunisia
Guillaume Bouleux, University of Saint Etienne | INSA-Lyon, France
Klaus Brinker, Hamm-Lippstadt University of Applied Sciences, Germany
Tolga Çakmak, Hacettepe University, Turkey
Manuel Campos Martínez, University of Murcia, Spain
Armand Castillejo, STMicroelectronics, France

Copyright Information

Identifying Key Factors in Right Ventricular Involvement in Ischaemic and Non-ischaemic Cardiomyopathy

Carlos Barroso-Moreno
Faculty of Biomedical and Health Sciences
Universidad Europea de Madrid
Madrid, Spain
email: 22014885@live.uem.es
0000-0002-1609-2267

Hector Espinos Morato
i3M Molecular Imaging
Universidad Europea de Valencia
Valencia, Spain
0000-0002-4089-1368

Enrique Puertas
School of Architecture, Engieering and Design
Universidad Europea de Madrid
Madrid, Spain
0000-0002-5115-1226

Juan José Beunza Nuin
Faculty of Biomedical and Health Sciences
Universidad Europea de Madrid
Madrid, Spain
0000-0001-8192-2952

José Vicente Monmeneu
Cardiac Magnetic Resonance
Exploraciones Radiológicas Especiales (ERESA)
Valencia, Spain
email: jmonmeneu@eresa.com

David Moratal
Center for Biomaterials and Tissue Engineering
Universitat Politècnica de Valencia
Valencia, Spain
0000-0002-2825-3646

María P. López-Lereu
Cardiac Magnetic Resonance
Exploraciones Radiológicas Especiales (ERESA)
Valencia, Spain
email: mplopezl@ascires.com

***Abstract*—Cardiomyopathy is a disease of the heart muscle that makes it harder for the heart to pump blood. Previous studies have focused on the left ventricle, but in recent years the relevance of the right ventricle has been the focus of current research. The aim is to determine those clinical and cardiac parameters that influence right ventricular involvement in ischaemic and non-ischaemic cardiomyopathy. The used database is composed of 56,447 subjects collected from 2008 to 2020 by ASCIRES Biomedical Group. The methodology is divided into two blocks: in the clinical aspect, decision trees are used to gain interpretability and in the technical aspect, Machine Learning (ML) is used for a greater degree of prediction. The results show the influence of the difference in aortic artery beat volume and vascular pulmonary volume as key factors, reaching an Area Under the Curve (AUC) of 92.3% using RapidMiner tool with decision trees algorithm. The conclusions demonstrate the ability to identify clinical variables of right ventricular involvement and consequently reduce the number of diagnostic tests and associated times in a situation of cardiomyopathy.**

***Index Terms*—Machine Learning; right ventricular involvement; Pulmonary Vascular Resistance; Cardiomyopathy.**

I. INTRODUCTION

Among cardiovascular diseases, ischaemic heart disease accounts for 16% of all deaths worldwide, rising from over 2 million deaths in 2000 to 8.9 million in 2019, and has become the disease attributed with the largest increase in deaths since 2000 [1]. Other cardiac conditions, such as non-ischaemic cardiomyopathy, arrhythmia, valvular heart disease, and heart failure are highly prevalent in developed countries and also cause high morbidity and mortality [2]. Due to the complexity and high prevalence of these diseases, a better understanding of the pathophysiology, as well as earlier diagnosis is of vital importance to increase the success rate of therapies, which is reflected in a reduced level of disability and lower mortality. To this end, for decades, all attention has been directed to the study of the left ventricle, making the right ventricle the "forgotten side of the heart". On the other hand, a direct extrapolation of the knowledge acquired about the physiology of the left side of the heart to the right side is not possible, as the normal right ventricle is anatomically and functionally different from the left ventricle. However, in recent years, advances in non-invasive cardiac imaging techniques have made it possible to discover the importance of the right ventricle in different cardiac diseases [3]. Therefore, there is a need for a better understanding of those factors that influence right ventricular dysfunction, given the accumulating evidence of their clinical relevance from both a symptomatic or diagnostic and prognostic perspective.

Numerous researchers have demonstrated the feasibility of applying Machine Learning (ML) algorithms in health studies to predict strokes [4], ICU patients with Covid-19 [5], prostate cancer [6] or acute coronary diseases syndrome [7] and others. The joint use of this type of algorithms with visualisation tools, such as Power BI is used in numerous areas [8] [9]

[10]. However, the diagnostic use of decision-making tools is still rare in the health sector. In recent years, these types of diagnostic aid tools have become popular. For example, a success story is the application of this type of tools in the private health sector in Finland, which has based its health system's decisions on data, identifying key factors [11].

Previous studies have identified influential variables in right ventricular compromise, such as: pulmonary arterial hypertension (PAH) associated with pressure overload [12]; diabetes, dyslipidemia [13], blood flow [14] and habits, such as smoking [15] and others.

The main objective of this project is to determine those clinical and cardiac parameters that influence the involvement of the right ventricle in ischemic and non-ischemic cardiomyopathy using ML techniques. To this end, predictive models capable of identifying patients with right ventricular dysfunction will be developed and the key parameters used by the models will be studied from the point of view of their clinical implication.

This paper is organized as follows. In Section 2 presents the details of dataset and it describes the methodology for Power BI and RapidMinner. Section 3 describes the results focusing on significant variables, distributions and decision trees . Finally, Section 4 presents the conclusions and directions for future work.

II. MATERIALS AND METHODS

The methodology used makes a comparison of the most common supervised classification algorithms in ML: Support Vector Machines (SVM) [16], decision tree [17], Random Forest [18] and neural networks [19]. To compare results, the following metrics have been used: precision, sensitivity, specificity and Area Under the Curve (AUC). This last variable has been used to evaluate the performance of binary classification models.

The data used for the study comes from the ASCIRES Biomedical Group database. This database has 56,447 records of variables collected from 2008 to 2020.

A. Software used

The research uses two software tools: Power BI and RapidMinner.

- Power BI is a data analytic service from Microsoft that provides interactive graphs focused on analytic intelligence to generate reports [20]. In the present research it allows easy visualisation of the database to automate the analysis.
- RapidMiner is a software for data analysis and data mining by chaining operations in a graphical environment [21]. Version 9.10.013 is used to obtain the results of the ML models.

B. Data preparation

At this point, clinical filters and patient labels are made according to age, gender, systole and diastole of the right ventricle. The volume of data cleaning by means of the filters means, that the initial database has 56,447 patients and 1,815 variables; after applying the filters, these are reduced to 12,083 patients and 120 variables. Of the latter subgroup, 7153 are labelled as unaffected and the target group is 4944 with right ventricular involvement.

The process applies logical cleaning, such as: (i) Removal of inconsistent data, such as the presence of letters in numerical values. (ii) Elimination of erroneous data, heights < 1 metre and > 2.3 metre or ages < 0 and > 120 years. (iii) Checking whether the numeric value zero represents such a value or is a null value (NULL). (iv) For having the same content as other variables, but with a different name, e.g., Vol.eyec.Ao for vol.lat. (v) For having all data set to 0; (vi) Transformation into international units of certain variables, such as wood units. (vii) For containing inconsistent data from the clinical perspective (outlayer).

An additional step is the elimination of variables with higher correlations, in order to avoid multicollinearity in our database. These steps are the following:

(i) Frac.reg.Ao.por.vol.lat to Ao.reg.Vol.beat.vol.dif with $\rho = 0,91$ (ii) Ao.reg.Vol.beat.vol.dif to Vol.reg.Pulm.dif.vol.lat with $\rho = -1$ (iii) Frac.reg.Ao.por.vol.lat to Vol.reg.Pulm.dif.vol.lat with $\rho = -0,92$ (iv) IMVI to MVI with $\rho = 0,93$ (v) NLVEDV to NLVESV with $\rho = 0,94$ (vi) Weight (kg) to S.Corp with $\rho = 0,95$ (vii) NRVSV to RVSV with $\rho = 0,95$ (viii) NLVSV to LVSV with $\rho = 0,95$ (ix) NRVEDV to RVEDV with $\rho = 0,95$ (x) RWT...relative.wall.thickness to RWT.2...relative.wall.thickness.pwd.sd with $\rho = 0,90$ (xi) LVEDV to LVESV with $\rho = 0,93$ (xii) NLVEDV to LVEDV with $\rho = 0,95$ (xiii) NLVEDv to LVESV with $\rho = 0,90$ (xiv) IVTSVD to RVESV with $\rho = 0,97$ (xv) NLVESV to LVESV with $\rho = 0,97$

C. Data labels

Aligned with the main objective of the research, the database is labelled to assess whether the patient has right ventricular involvement. For the identification of these patients, the consensus tables specified by the European Society of Cardiology [22] establishes ranges of variables (age, gender, systoles, diastole and others) to identify RV involvement. These values are adapted in Table I. The normal values of RV systolic and diastolic parameters vary according to age and gender. The standard of normal values used for the recognition of impairment is used with a similar range in current papers, such as that of the researchers Petersen et al. (2019) [23].

The absolute values of end-systolic volume (ESV), end-diastolic volume (EDV) and body mass have been used, whose values are provided automatically within the framework of clinical tests. Clinics automatically provide based on clinical evidence frameworks.

Systolic volume (SV) is calculated by the difference between EDV and ESV; Additionally, ejection fraction (EF) is calculated as SV/VDE. Sex, body surface area (BSA), and age are independent predictors of several RV parameters, as suggested by previous well-established studies [24]. The

TABLE I
RIGHT VENTRICLE LABELS. STANDARD RANGES BY RV VOLUMES, SYSTOLIC FUNCTION AND MASS BY AGE INTERVAL (95% CONFIDENCE INTERVAL). ADAPTED FROM MACEIRA ET AL. (2006) [22].

Right Ventricule labels						
Age (years)	*20-29*	*30-39*	*40-49*	*50-59*	*60-69*	*70-79*
Males						
Absolute values						
1-1 EDV (mL) SD 25.4	(127,227)	(121,221)	(116,216)	(111,210)	(105,205)	(100,200)
ESV (mL) SD 15.2	(38,98)	(34,94)	(29,89)	(25,85)	(20,80)	(16,76)
SV (mL) SD 17.4	(74,143)	(74,142)	(73,141)	(72,140)	(71,139)	(70,138)
EF (%) SD 6.5	(48,74)	(50,76)	(52,77)	(53,79)	(55,81)	(57,83)
Mass (g) SD 14.4	(42,99)	(40,97)	(39,95)	(37,94)	(35,92)	(33,90)
Normalized to BSA						
EDV/BSA (mL//m^(2)) SD 11.7	(68,114)	(65,111)	(62,108)	(59,105)	(56,101)	(52,98)
ESV/BSA (mL//m^(2)) SD 7.4	(21,50)	(18,47)	(16,45)	(13,42)	(11,40)	(8,37)
Females						
Absolute values						
1-1 EDV (mL) SD 21.6	(100,184)	(94,178)	(87,172)	(81,166)	(75,160)	(69,153)
ESV (mL) SD 13.3	(29,82)	(25,77)	(20,72)	(15,68)	(11,63)	(6,58)
SV (mL) SD 13.1	(61,112)	(59,111)	(58,109)	(56,108)	(55,106)	(53,105)
EF (%) SD 6	(49,73)	(51,75)	(53,77)	(55,79)	(57,81)	(59,83)
Mass (g) SD 10.6	(33,74)	(31,72)	(28,70)	(26,68)	(24,66)	(22,63)
Normalized to BSA						
1-1 EDV/BSA (mL//m^(2)) SD 9.4	(65,102)	(61,98)	(57,94)	(53,90)	(49,86)	(45,82)
ESV/BSA (mL//m^(2)) SD 6.6	(20,45)	(17,43)	(14,40)	(11,37)	(8,34)	(6,32)

standardised values of EDV/BSA and ESV/BSA are obtained from these variables.

The filtered database uses functions to apply binary labeling of patients: Normal or Abnormal (RV involvement). As an example, a 62-year-old male patient with an EDV of 211 EDV (mL) is labeled as abnormal (RV impairment) because he is not within the range of (105,205) set in the parameters of Table I. If the same patient has an EDV of 204 EDV (mL), the patient is considered to have no RV involvement (normal), if the patient also meets the other variables in their corresponding ranges.

The database input and output variables are described in Table II. The $\mathbf{\bar{X}}$ is the average; SD is the standard deviation; Max is the maximum value and Min is the minimum value. The database contains 12,083 patients, of which 76.6% ($n = 9260$) are men and 23.4% ($n = 2823$) are women. The mean age is 62.49 years with a standard deviation of 14.1. The average body mass index (BMI) is 27.87 ($SD = 4.69$), the formula is $BMI = Weight(kg)/[Height(m)]^2$. According to the BSA, the average is 18.04 ($SD = 2.8$).

The output result corresponds to the classification of RV involvement with patients without RV involvement (normal) in a percentage of 59.1% ($n = 7139$) and with RV involvement (abnormal) in a percentage of 40.9% ($n = 4944$). Missing data is 0 because these records are removed in the preprocessing step so as not to distort the output of the ML algorithms in later steps.

In the first tests, the variables in the table are eliminated due to their direct relationship in the table calculations. These are discarded, as detailed in Table III: right ventricular systolic volume (RVSV), left ventricular systolic volume (LVSV), right ventricular end-diastolic volume (RVEDV), left ventricular end-diastolic volume (LVEDV), right ventricular end-systolic volume (RVESV), left ventricular end-systolic volume (LVESV). However, they are reintroduced in the prediction tests, as there is no clear correlation with the output variable and they are not detected as primary variables for predicting RV involvement.

TABLE II
INPUT VARIABLES FOR OUTPUT VARIABLE LABEL.

Input				
Variable	*Categories*	*n*	*%*	*Missing*
Gender	Males	9260	76,6	0
	Females	2823	23,4	0
	X̄	*SD*	*Min*	*Max*
Age	62,49	14,1	20	95
BMI	27,87	4,69	13,1	71,4
BSA	18,04	2,8	4,22	53,13
EDV/BSA	70,94	28,34	12,0	403,3
ESV/BSA	33,86	21,03	0,9	251,9
Output				
Variable	*Categories*	*n*	*%*	*Missing*
DV involvement	Normal	7139	59,1	0
	Abnormal	4944	40,9	0

III. RESULTS

After data preparation with a single model, the samples of the training set (n-train) and test sets (n-test) are 8458 and 3625, respectively. These patients are changed by cross-validation in the tests.

The results of the algorithms are shown in the table IV, which are Accuracy (ACC), Sensitivity (SE), Specificity (SP), Positive Predictive Value (PPV), Negative Predictive Value (NPV) and Area Under the Curve (AUC).

TABLE III
VARIABLES ELIMINATED BY INDIRECT USE IN THE OUTPUT VARIABLE.

	Elimnated variables			
	$\overline{X}$	*SD*	*Min*	*Max*
RV.DTD	32,78	7,33	6	100
RVSV	70,73	27,24	4	250
LVSV	75,05	24,16	8	200
RVEDV	135,5	55,69	23	500
LVEDV	196,89	75,97	33	600
RVESV	64,68	40,56	2	437
LVESV	122,08	69,62	15	500

The algorithm with the highest AUC in RapidMiner is performed with XGBoost ($AUC = 0.87$), although the difference is small with the neural network algorithm ($AUC = 0.85$), determining a predominance of neural networks as predictors RV involvement. Although Random Forest algorithms have a ($AUC = 0.82$), and the lowest of the results is from Support Vector Machine (SVM) with a AUC =0.79. The results are quite promising, as it means that we can predict almost with an accuracy of 9 out of 10 patients performing the clinical tests whether they have RV involvement.

However, direct comparisons of the algorithms should be treated with caution. On the one hand, XGBoost provides the best results of the study but the complexity of ML algorithms makes them "black boxes". On the other hand, decision trees obtain inferior results but allow for great interpretability of the results gaining a value-added clinical perspective. The size of the database and the optimisation of the algorithms used allow low processing times of less than 3 minutes with a MacBook Pro 2.6 GHz Intel Core i7 2.6 GHz 6-core computer with 6 GB 2400 MHz DDR4 memory.

TABLE IV
COMPARISON AND EVALUATION OF DIFFERENT ALGORITHMS IN RAPIDMINNER.

Algorithms	*ACC*	*SE*	*SP*	*NPV*	*AUC*
SVM	0.76	0.76	0.76	0.77	0.79
Random Forest	0.78	0.78	0.79	0.78	0.82
Neural Network	0.78	0.78	0.79	0.77	0.85
XGBoost	0.82	0.78	0.80	0.77	0.87

The Table V shows the main variables with the greatest weight detect in the ML algorithms, whose selection is automatic as key risk predictors. The descriptive analysis of the variables are gender in male with 76.6% ($n = 9260$) and female 23,4% ($n = 2823$). This is logical as it is a necessary variable for patient labelling. Dyslipidemia is present in 51.6% ($n = 6231$) of patients, hypertension has a 59.0%($n = 7133$), diabetes has a low presence with a prevalence rate of 0.7% ($n = 85$), however, type II diabetes (low involvement) has a high incidence with 67.3% ($n = 8131$). The rest of the patients of each typology do not present the pathology. Smoking patients are high with 72.8% ($n = 8800$). Stent implantation is high with 82.2% ($n = 9937$) because the database is made up of patients who go to the cardiologist and suffer from some kind of affectation or signs of affectation of the heart. The stress study as a medical test is performed in 48.5% of patients ($n = 5857$), although the use of this aggressive test is decreasing every year.

In reference to the numerical variables, the most relevant is RVP [Wood] with $\overline{X} = 14,99$ ($SD = 1,27$). El aortic arch is $\overline{X} = 25,71$ ($SD = 3,68$), height is $\overline{X} = 167,86$ ($SD = 9,13$), Weight (kg) is $\overline{X} = 78,63$ ($SD = 14,96$). The Regurgitate Volume of the Aorta Artery (Reg.Vol.Ao.) is available for $\overline{X} = 4,31$ ($SD = 24,27$). The Diameter of the Aorta with Pulmonary Artery (DoA.PA) is $\overline{X} = 0,96$ ($SD = 0,25$). The Descending Thoracic Aorta (Desc.T.Ao) es de $\overline{X} = 24,43$ ($SD = 3,85$). Sinus pressure (Sinus.P) is $\overline{X} = 34,1$ ($SD = 4,77$). Posterior Wall in Diastole (PWD) has an average of $\overline{X} = 8,63$ ($SD = 2,75$). Finally, the ratio of diastolic volumes (diastolic RV) are of $\overline{X} = 1,59$ ($SD = 0,71$).

TABLE V
ANALYSIS OF THE VARIABLES WITH THE GREATEST WEIGHT IN THE ML ALGORITHM IN THE ASCIRES BIOMEDICAL GROUP DATABASE.

	Main variables			
Variable	*Categories*	*n*	*%*	*Miss.*
Gender	Males	9260	76,6	0
	Females	2823	23,4	0
Dyslipidemia	Yes	6231	51,6	0
	No	5852	48,4	0
Hypertension	Yes	7133	59,0	0
	No	4950	41,0	0
Diabetes	Yes	85	0,7	0
	No	11998	99,3	0
Diabetes II	Yes	8131	67,3	0
	No	3952	32,7	0
Smoker	Yes	8800	72,8	0
	No	3283	27,2	0
Stent	Yes	9937	82,2	0
	No	2146	17,8	0
Stress study	Yes	5857	48,5	0
	No	6226	51,5	0
	$\overline{X}$	*SD*	*Min*	*Max*
PVR [wood]	14,99	1,27	0,49	18,97
Aortic.Arch	25,71	3,68	2	60
Height (cm)	167,86	9,13	131	205
Weight (kg)	78,63	14,96	35	187
Vol.reg.Ao. dif.vol.lat	4,31	24,27	-247	305
DoA.PA	0,96	0,25	0,01	5,01
Desc.T.Ao	24,43	3,85	2	63
Sinus.P	34,1	4,77	9	83
PWD	8,63	2,75	1	125
diastolic RV	1,59	0,71	0,17	10,88

A. *Pulmonary Vascular Resistance (PVR)*

Pulmonary Vascular Resistance (PVR) is the mean pressure drop from the main pulmonary artery to the divided left atrium. The units of measurement are Wood's units, which arise from the equivalence one Wood's unit = 80·s·cm^{-5}.

PVR is defined by the Swan-Ganz catheter from a central vein [25], the formula is:

$$PVR = 80 * (PAP - CEP) * CO, \quad (1)$$

therefore depend on Pulmonary Arterial Pressure (PAP) in mmHg units; the Capillary Locking Pressure (CEP) in mmHg units; and Cardiac Output (CO) in l/min units. The normal value for a subject is 1-2 [Wood], but this increases with age, as determined by the correlation table and Vizza et al. (2022) [26], increasing by 0.2 Wood in subjects over 50 years of age.

The database provided uses an estimation model [27]:

$$PVR[Wood] = 19.38-(4.62*Ln(PAAV)-(0.08*RVEF)) \tag{2}$$

where PAAV are in centimeter per second and RVEF in percentage.

B. Distributions

The raw database model not shown in the results of the paper has an AUC value of 98%, the reason is that most of the patients in the database do not have RV involvement. This caused the algorithms to estimate a normal value by default and thus be right in most cases (sensitivity almost 0). But after data filtering, the database is balanced, preventing this negative event in the analysis of the observations. Another noteworthy aspect of the distributions is the elimination of data in the filtering that are empty in any of the variables, which may cause a bias by eliminating patients who intrinsically did not complete the data due to some particularity. As the volume is high, this is not an impediment to the analysis and the algorithms yield better results.

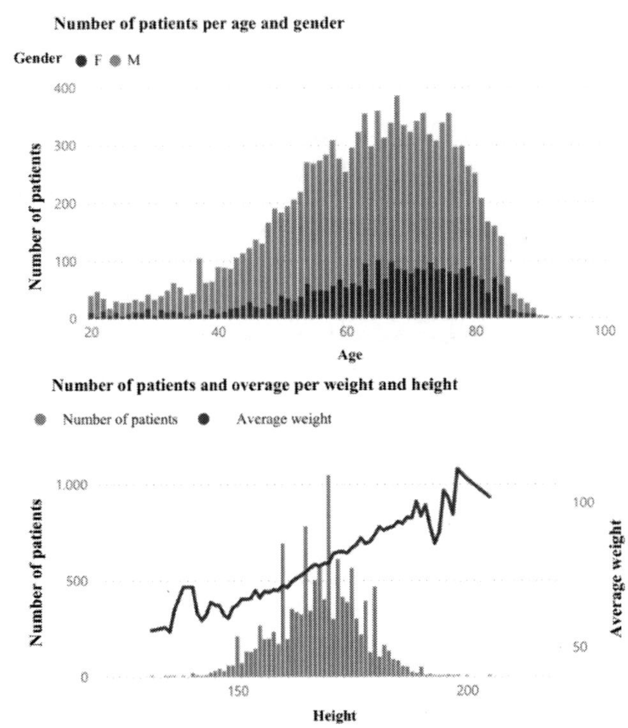

Fig. 1. Multivariate histograms of number of patients, (top) age as a function of gender, (bottom) height as a function of mean weight.

The use of Power BI makes it possible to visualise these and many other particularities quickly, avoiding bias or inconsistencies in the data. The Figure 1 shows the histogram of the number of patients by age and gender, quickly showing how an increase in age means a higher incidence, as well as a predominantly male gender. The 60 to 80 year age range accounts for almost half of the RV affectations. In reference to height and weight, there is a clear correlation between the increase in height and weight, which is why BMI is used as a predictive variable in the labelling of the database.

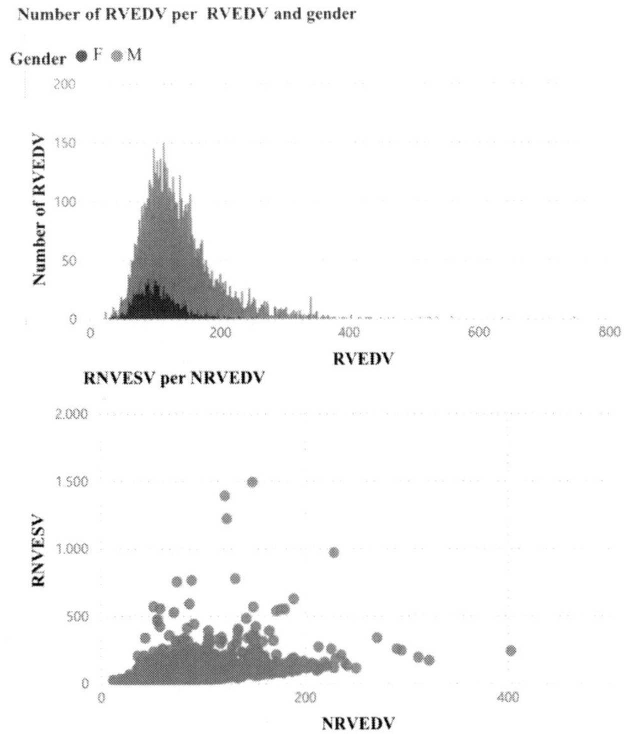

Fig. 2. Multivariate histogram of the number of patients, (top) RVEDV by gender, and scatter plot of NRVEDV with RNVESV.

Figure 2 shows the multivariate analysis of RVEDV according to gender and number of patients, in the analysis it can be seen that the female anatomy has a smaller volume than the male anatomy. This situation is contemplated in the labelling of RV involvement, so this differentiation from previous studies is correct. Regarding the lower figure, it represents the dispersion in relation to NRVEDV with ICTSVD, this analysis allows the detection of outliers to eliminate patients with highly distorted values. These graphs allowed to establish minima and maxima in the variables without interfering negatively in the algorithms. The interactive graphs allow for easy filtering of the outliers in order to reach a consensus with the doctors on their discarding and clinical criteria with a range limit for each variable. This process is repeated for each of the variables used in the labelling of the data.

According to PVR [Wood], Figure 3, which represents the

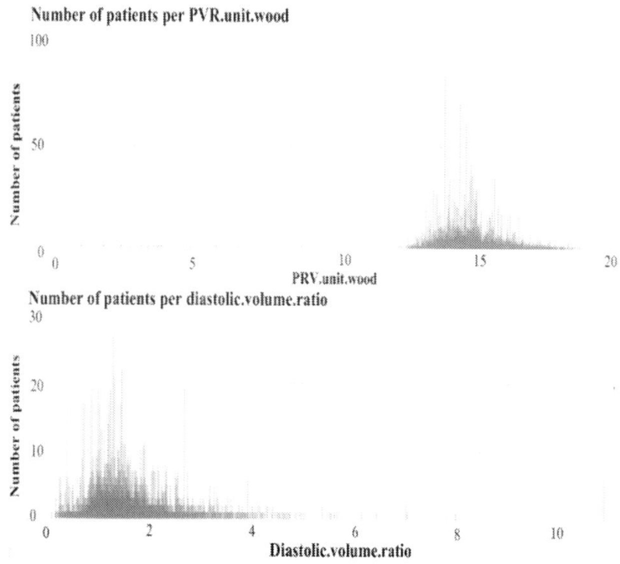

Fig. 3. Multivariate histograms of the number of patients, (top) as a function of PVR [Wood], (bottom) as a function of diastolic volume ratio.

data distributions concentrated in the ranges of 12 to 18 [Wood], establishing a valid variable quality to be considered relevant in the research. This concentration allows us to discard outlayer patients (higher than 6) to avoid confusion of the algorithms in the predictions.

C. Decision trees

The decision tree has a couple of advantages, the first, as already mentioned, have and easily interpretability for clinicians in decision making, the second is easy data manipulation, and the third is speed of execution and design. The disadvantage is an algorithm with low predictive power.

A pruning is performed on the best decision tree, however, if it is too high we can produce an overfitting. The hyperparameters used are adjusted for avoiding overfitting. Overfitting is not observed.

According to the algorithm, the patient can be classified with an AUC (91.2%) for RV involvement, based on the interpretation of the provided tree, Figure 4:

- RV involvement (Abnormal): If PVR.en.units.wood > 15.221 + Height > 1.40m + DoA.PA > 0.087 + Aortic.arch > 14.
- No RV involvement (normal): If PVR.en.units.wood ¡ 15.221 and Ao.reg.Vol.beat.vol.dif >-90.500.

With this philosophy, decision trees are generated to generate visualisations that follow the branches to generate those visualisations on the medical side. On the technical side we use more complex but less visual algorithms to fit higher quality predictor data. Another important variable in the decision tree is Aortic Regurgitation Volume by Beat-Volume Difference (Ao.reg.Vol.beat.vol.dif).

The following tree would be a similar interpretation, if we remove the variable PVR.in.wood.units, the variable reason.of.diastolic.volumes stands out, but the accuracy is reduced to 66.6% in the decision tree and 78.03% in the Gradient Boosted Trees.

If we include the variables gender and age, which were not initially included because they are used to classify the DV affectations, the results do not change. However, if we prune the tree, the variables age and gender are included, but they are not decisive, and a test with grouping by decades of age is carried out to ensure that they win.

The left ventricular involvement variables have been included; the algorithm does not identify them as a key element for right ventricular involvement. Although it is true that there is joint involvement and influence.

Previous studies with ML algorithms in cardiology achieve predictions greater than 90%, as they focus on achieving the best results, not on their interpretability [28]. It is interesting to see that our study achieves close values with interpretability by combining both tools (RapidMiner and Power BI). The results obtained are in line with recent research, which states that neural networks are the most predictive of cardiac parameters [29] [30].

This investigation has some relevant points. The main one is the creation of a tool to support clinical diagnosis that cardiologists can use for the prescription of new tests and a more detailed follow-up, similar to previous experiences already carried out [4] [31]. A second point is the creation of a visual interface, which allows dynamic monitoring, which facilitates dynamic interpretation, generating reports of high statistical value. It is worth exploring new algorithms to improve the interpretation of those key factors in the involvement of the right ventricle, thus improving the possible diagnosis. Tests could be carried out with different databases to create an algorithm that is robust enough to be able to limit any bias that the database used may contain. On the other hand, the authors aim to create a standardized protocol of measurements and tests that is carried out in daily clinical practice. The benefits of data analysis in cardiology using these types of techniques are evident, allowing them to increase the quality of diagnosis, prognosis and therapy.

IV. CONCLUSION

The ML algorithms and decision trees described, demonstrate the ability to identify the variables of greatest weight in right ventricular involvement with a reduced number of parameters. As a result, the number of diagnostic tests and their associated times can be reduced, allowing faster intervention in ischaemic and non-ischaemic cardiomyopathy. Therefore, the main objective of determining the clinical parameters that influence the involvement of the right ventricle in ischaemic and non-ischaemic cardiomyopathy is fulfilled.

The results show the influence of pulmonary vascular resistance and aortic artery beat volume difference as key factors, reaching an AUC of 87,7% with XGBoost in decision trees. Other variables with height (BMI), DoA.PA and Ao.reg.Vol.beat.vol.dif stand out as variables with a high weight in the predictions.

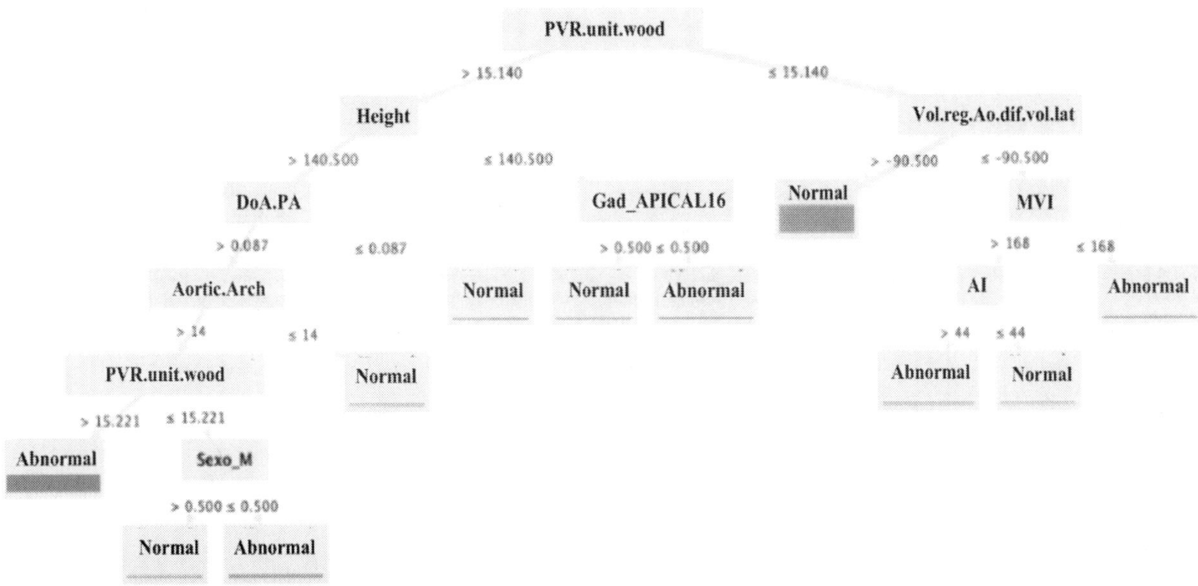

Fig. 4. Decision tree pruned in RapidMiner.

Author contributions

CB designed the article, performed the analysis in Power BI and drafted the text. HE co-managed the development of the article and obtained the database permissions. EP performed the ML analysis with RapidMiner. JJB engineering and medical design consultancy with descriptive analysis of the database. JVM interpretation of medical results. DM engineering design consultancy. MPL interpretation of medical results.

Declaration of Competing Interest

The authors declare that they have no known competing financial interests or personal relationships that could have appeared to influence the work reported in this paper.

Acknowledgements

The research is funded by the European University through the project "Identification of factors influencing right ventricular involvement in ischaemic and non-ischaemic cardiomyopathy" through the project 2022/UEM18. The authors of this study wish to express their gratitude to ASCIRES Biomedical Group for the confidentiality agreement reached with the European University; Silvia Ruiz-España (Universitat Politècnica de València) for the documentation compiled from the database; to the advice of the interdisciplinary working group of Machine Learning Health-UEM; as well as to many other researchers for being a source of inspiration in the convergence of technology applied to health.

REFERENCES

[1] O. W. Health, "The top 10 causes of death in the world." OWH, 2020. [Online]. Available: https://www.who.int/es/news-room/fact-sheets/detail/the-top-10-causes-of-death

[2] E. Boot, M. S. Ekker, J. Putaala, S. Kittner, F. E. De Leeuw, and A. M. Tuladhar, "Ischaemic stroke in young adults: a global perspective," *Journal of Neurology, Neurosurgery & Psychiatry*, vol. 91, no. 4, pp. 411–417, 2020.

[3] L. Cavigli, M. Focardi, M. Cameli, G. E. Mandoli, S. Mondillo, and F. D'Ascenzi, "The right ventricle in "left-sided" cardiomyopathies: the dark side of the moon," *Trends in Cardiovascular Medicine*, vol. 31, no. 8, pp. 476–484, 2021.

[4] J. J. Beunza, E. Puertas, E. García-Ovejero, F. Villalba, E. Condes, G. Koleva, C. Hurtado, and M. F. Landecho, "Comparison of machine learning algorithms for clinical event prediction (risk of coronary heart disease)," *Journal of biomedical informatics*, vol. 97, p. 103257, 2019.

[5] S. Martínez-Agüero, A. G. Marques, I. Mora-Jiménez, J. Álvarez Rodríguez, and C. Soguero-Ruiz, "Data and network analytics for covid-19 icu patients: A case study for a spanish hospital," *IEEE Journal of Biomedical and Health Informatics*, vol. 25, no. 12, pp. 4340–4353, 2021.

[6] R. Cuocolo, M. B. Cipullo, A. Stanzione, L. Ugga, V. Romeo, L. Radice, A. Brunetti, and M. Imbriaco, "Machine learning applications in prostate cancer magnetic resonance imaging," *European radiology experimental*, vol. 3, no. 1, pp. 1–8, 2019.

[7] A. García-García, I. Prieto-Egido, A. Guerrero-Curieses, J. R. Feijoo-Martínez, S. Muñoz-Romero, S. Fernández-Manzano, P. J. Flores-Blanco, J. L. Rojo-Álvarez, and A. Martínez-Fernández, "Data science analysis and profile representation applied to secondary prevention of acute coronary syndrome," *IEEE Access*, vol. 9, pp. 78 607–78 620, 2021.

[8] D. Andriansyah and L. Nulhakim, "The application of power business intelligence in analyzing the availability of rental units," in *Journal of Physics: Conference Series*, vol. 1641, no. 1. IOP Publishing, 2020, p. 012019.

[9] M. Mariani, R. Baggio, M. Fuchs, and W. Höepken, "Business intelligence and big data in hospitality and tourism: a systematic literature review," *International Journal of Contemporary Hospitality Management*, vol. 30, no. 12, pp. 3514–3554, 2018.

[10] A. Sánchez-Ferrer, H. Pérez-Mendoza, and P. Shiguihara-Juárez, "Data visualization in dashboards through virtual try-on technology in fashion industry," in *2019 IEEE Colombian Conference on Applications in Computational Intelligence (ColCACI)*. IEEE, 2019, pp. 1–6.

[11] M. Ratia, J. Myllärniemi, and N. Helander, "The new era of business intelligence: Big data potential in the private health care value creation," *Meditari Accountancy Research*, vol. 26, no. 3, pp. 531–546, 2018.

[12] C. R. Greyson, "Ventrículo derecho y circulación pulmonar: conceptos básicos," *Revista española de cardiología*, vol. 63, no. 1, pp. 81–95, 2010.

[13] J. Sanz, D. Sánchez-Quintana, E. Bossone, H. J. Bogaard, and R. Naeije, "Anatomy, function, and dysfunction of the right ventricle: Jacc state-of-the-art review," *Journal of the American College of Cardiology*, vol. 73, no. 12, pp. 1463–1482, 2019.

[14] S. Wang, H. Wang, M. Ng, Y. Tada, G. Pontone, J. Urmeneta, I. Saeed, H. Patel, C. Mariager, J. V. Monmeneu-Menadas *et al.*, "Quantification of myocardial blood flow using stress cardiac magnetic resonance for the detection of coronary artery disease," *European Heart Journal-Cardiovascular Imaging*, vol. 24, no. Supplement_1, pp. jead119–375, 2023.

[15] J. M. Oakes, J. Xu, T. M. Morris, N. D. Fried, C. S. Pearson, T. D. Lobell, N. W. Gilpin, E. Lazartigues, J. Gardner, and X. Yue, "Effects of chronic nicotine inhalation on systemic and pulmonary blood pressure and right ventricular remodeling in mice," *Hypertension*, vol. 75, no. 5, pp. 1305–1314, 2020.

[16] C. Cortes and V. Vapnik, "Support-vector networks," *Machine learning*, vol. 20, pp. 273–297, 1995.

[17] S. K. Murthy, "Automatic construction of decision trees from data: A multi-disciplinary survey," *Data mining and knowledge discovery*, vol. 2, pp. 345–389, 1998.

[18] T. Ho-Kam, "Random decision forests," in *Proceedings of 3rd international conference on document analysis and recognition*, vol. 1. IEEE, 1995, pp. 278–282.

[19] T. Chen and C. Guestrin, "Xgboost: A scalable tree boosting system," in *Proceedings of the 22nd acm sigkdd international conference on knowledge discovery and data mining*, 2016, pp. 785–794.

[20] T. Lachev and E. Price, *Applied Microsoft Power BI Bring your data to life!* Prologika Press, 2018.

[21] V. Kotu and B. Deshpande, *Predictive analytics and data mining: concepts and practice with rapidminer*. Morgan Kaufmann, 2014.

[22] A. M. Maceira, S. K. Prasad, M. Khan, and D. J. Pennell, "Reference right ventricular systolic and diastolic function normalized to age, gender and body surface area from steady-state free precession cardiovascular magnetic resonance," *European heart journal*, vol. 27, no. 23, pp. 2879–2888, 2006.

[23] S. E. Petersen, M. Y. Khanji, S. Plein, P. Lancellotti, and C. Bucciarelli-Ducci, "European association of cardiovascular imaging expert consensus paper: a comprehensive review of cardiovascular magnetic resonance normal values of cardiac chamber size and aortic root in adults and recommendations for grading severity," *European Heart Journal-Cardiovascular Imaging*, vol. 20, no. 12, pp. 1321–1331, 2019.

[24] J. M. Gardin, M. K. Rohan, D. Davidson, A. Dabestani, M. Sklansky, R. Garcia, M. L. Knoll, D. White, S. K. Gardin, and W. L. Henry, "Doppler transmitral flow velocity parameters: relationship between age, body surface area, blood pressure and gender in normal subjects," *American journal of noninvasive cardiology*, vol. 1, no. 1, pp. 3–10, 1987.

[25] J. Swan-Harold, W. Ganz, J. Forrester, H. Marcus, G. Diamond, and D. Chonette, "Catheterization of the heart in man with use of a flow-directed balloon-tipped catheter," *New England Journal of Medicine*, vol. 283, no. 9, pp. 447–451, 1970.

[26] C. D. Vizza, I. M. Lang, R. Badagliacca, R. L. Benza, S. Rosenkranz, R. J. White, Y. Adir, A. K. Andreassen, V. Balasubramanian, S. Bartolome *et al.*, "Aggressive afterload lowering to improve the right ventricle: a new target for medical therapy in pulmonary arterial hypertension?" *American journal of respiratory and critical care medicine*, vol. 205, no. 7, pp. 751–760, 2022.

[27] A. García-Alvarez, L. Fernandez-Friera, J. G. Mirelis, S. Sawit, A. Nair, J. Kallman, V. Fuster, and J. Sanz, "Non-invasive estimation of pulmonary vascular resistance with cardiac magnetic resonance," *European heart journal*, vol. 32, no. 19, pp. 2438–2445, 2011.

[28] T. Smole, B. Žunkovič, M. Pičulin, E. Kokalj, M. Robnik-Šikonja, M. Kukar, D. I. Fotiadis, V. Pezoulas, N. S. Tachos, F. Barlocco *et al.*, "A machine learning-based risk stratification model for ventricular tachycardia and heart failure in hypertrophic cardiomyopathy," *Computers in biology and medicine*, vol. 135, p. 104648, 2021.

[29] P. Revuelta-Zamorano, A. Sánchez, J. L. Rojo-Álvarez, J. Álvarez-Rodríguez, J. Ramos-López, and C. Soguero-Ruiz, "Prediction of healthcare associated infections in an intensive care unit using machine learning and big data tools," in *XIV Mediterranean Conference on Medical and Biological Engineering and Computing 2016: MEDICON 2016, March 31st-April 2nd 2016, Paphos, Cyprus*. Springer, 2016, pp. 840–845.

[30] R. Garcia Carretero, L. Vigil-Medina, O. Barquero-Perez, I. Mora-Jimenez, C. Soguero-Ruiz, and J. Ramos-Lopez, "Machine learning approaches to constructing predictive models of vitamin d deficiency in a hypertensive population: a comparative study," *Informatics for Health and Social Care*, vol. 46, no. 4, pp. 355–369, 2021.

[31] A. Hernández-Casillas, . Del-Canto, S. Ruiz-España, M. P. López-Lereu, J. V. Monmeneu, and D. Moratal, "Detection and classification of myocardial infarction transmurality using cardiac mr image analysis and machine learning algorithms," in *2022 44th Annual International Conference of the IEEE Engineering in Medicine & Biology Society (EMBC)*. IEEE, 2022, pp. 1686–1689.

Biomechanical Perspective on Effect of Angle in Arm Swing Movement on Vertical Ground Reaction Force for Gait Improvement

Sota Miura
Kochi University of Technology
Tosayamada, Kami, Kochi, 782-8502, Japan
e-mail: ssanpu98@gmail.com

Kyoko Shibata
Kochi University of Technology
Tosayamada, Kami, Kochi, 782-8502, Japan
email: shibata.kyoko@kochi-tech.ac.jp

Abstract— **This study focuses on arm swinging movements in gait for active self-healthcare. However, most of the research on arm swing movements is clinical approach and biomechanical mechanisms of why these movements are effective are not well understood. Therefore, this study aims to biomechanically elucidate the coupled process of arm swing and lower limb movement. This coupled process can lead to an effective gait according to his/her gait at the time. In this paper, we focus on the bimodality of vertical ground reaction force, which is also used for gait evaluation in clinical practice. The purpose of this paper is to experimentally clarify how the bimodality of vertical ground reaction force varies with arm swing, and to determine mechanistically what coupled processes are responsible for this variation. The experiment is carried out with four volunteers. In the experiment, each volunteer performs six different gait conditions. In these experiments, the ground reaction forces and mechanical parameters of each body segment are measured. The results of the analysis showed that an increase in the angle of lateral pelvic tilt and the vertical ground reaction force increased in the valleys and decreased in the part of late peaks, as the angle of arm swing increased. The angle of lateral pelvic tilt also showed a change in value at the same time as the late peak of the vertical ground reaction force. This suggests that the bimodality of vertical ground reaction force is related to the pelvic lateral tilt movement induced by arm swing.**

Keywords - Biomechanics; Pelvic angle; Arm movement; ground reaction force .

I. INTRODUCTION

Recent advances in medical care have led to the aging of the population in many developed countries [1]. In Japan, according to the 2021 Simple Life Chart published by the Ministry of Health, Labour and Welfare, the average life expectancy for men is 81.47 years and 87.57 years for women, and the percentage of people reaching 90 years of age is 3.9 times higher for men and 3.3 times higher for women compared to 1980[2]. These data suggest that society will continue to age further. Under these circumstances, the social issue is to reduce the gap between average life expectancy and healthy life expectancy. Walking is often viewed as the easiest way to solve this problem and as it has a low risk of disability even for people with knee or back diseases [3]. Therefore, walking has been studied from various approaches. Research on the effects of arm swinging has revealed changes in lower limb muscle activity [4], increased stride length [5][6], increased maximum walking speed [7], and improved stability in walking [8]. However, these studies are clinical approach, and it is not known why arm swing movements are effective. Consequently, depending on an individual's gait, changing the arm swing movement may not result in an effective gait, such as no change in walking speed or stride length, or conversely, a worsening of walking speed or stride length. Therefore, research is conducted from the point of view of biomechanics to clarify the factors that contribute to the effectiveness of walking. Most biomechanical studies that focus on arm swing have discussed the stability and symmetry of gait [9], but few have discussed how arm swing leads to an increase in walking speed and stride length. Hence, this study aims to biomechanically elucidate the coupled process of arm swing and lower limb movement. This coupled process can lead to an effective gait according to his/her gait at the time.

As a first step, this article focuses on the vertical ground reaction force, which is one of the most important indices in the evaluation of walking behavior [10] and is related to the propulsive force of walking. It is clinically known that the vertical ground reaction force is bimodal in normal gait and the bimodality of force changes with differences in gait. For example, the difference between the peaks and valleys of the bimodality of vertical ground reaction force increases with increasing walking speed [11], and the early peaks become larger, and the valleys and late peaks become smaller with increasing stride length [12], indicating a close relationship between walking motion and the bimodality of vertical ground reaction force. Additionally, the bimodality of vertical ground reaction force has been reported to be lost with aging due to a decrease in walking speed [13]. Therefore, the purpose of this report is to experimentally clarify how the bimodality of vertical ground reaction force is affected by the arm swing movement and to clarify the coupling between the arm swing movement and the vertical ground reaction force by observing the mechanical parameters. In the experiment, an optical motion capture system and a ground reaction force meter are used to measure and analyze the motion of each segment and the values of the ground reaction force induced by the swing of the arm setting the angle of arm swing condition.

Section II describes the conditions and methods of the arm swing experiment, Section III presents the results of the analysis, Section IV discusses the results of the analysis, and Section V summarizes this paper and discusses future developments.

II. Experimental Method

The purpose of this experiment is to measure the mechanical parameters and the vertical ground reaction force of each segment during gait to clarify the effect of adjusting the angle of arm swing on the vertical ground reaction force values. This experiment will focus on steps 5-7, which is the steady-state gait. The purpose and contents of this study were explained to four volunteers (age: 22.5±0.5 [years], height: 1.70±0.03 [m], weight: 66±2 [kg]) who gave oral and written consent. In addition, this study was approved by the Ethical Review Committee. There are six gait conditions: (a) walking natural without awareness of arm swing, walking with arms attached to the body (b) without swinging arms, (c) backward arm swing, (d) large backward arm swing, (e) forward arm swing, and (f) large forward arm swing, with five trials in each condition. In the case of a large arm swing, the upper body motion would differ between backward and forward swings, the effect of the upper body on the lower limbs would also be different. For the large arm swinging movement, the volunteer is instructed to raise his arms to a level parallel to the ground within a comfortable range. In addition to arm swing, we do not give any instructions on stride length or walking speed, considering the ease with which the volunteer can walk. In all walking conditions, except for natural walking and walking without swinging arms, volunteers practice walking five times before each trial to familiarize themselves with the walking motion.

To obtain a large number of mechanical parameters, such as muscle activity and joint angles, from body data, the musculoskeletal modeling simulation AnyBody (AnyBody Technology), one of the analysis software, was used. The 3D positional coordinates of the markers are measured by taking pictures using the MAC3D system (Motion Analysis Corporation), which is an optical motion capture system. Twelve cameras with a sampling frequency of 100 [Hz] are used. The measured marker positions are denoised with a low-pass filter of 6 [Hz]. Three force plates (TF-4060 and TF-6090 manufactured by Tec-Gihan) were used to measure reaction force values, with the right foot, left foot, and right foot moving on one force plate in the order of 5 to 7th step of the experimental volunteer, respectively. After the above measurements, a total of 119 data, excluding one trial of the forward arm swing condition for Volunteer A, whose data were found to be incomplete, are used as input data for AnyBody analysis.

III. Results

First, the results of the vertical ground reaction force values are shown. Figure 1 shows the vertical ground reaction force values for each randomly selected condition for volunteer D, representing Volunteers B~D who showed similar tendencies, and Volunteer A, who showed different tendencies from the other collaborators. The abscissa axis indicates the time when the right foot lands on the ground as 0[%] and the time when the right foot leaves the ground as 100[%].In natural walking, an early peak appears in the first half of the stance phase (around 20[%]), a valley in the second half of the stance phase (around 40[%]), and a late peak in the terminal stance phase (around 80[%]).

Table I summarizes the mean values of the peaks and troughs of the vertical ground reaction force for each volunteer and condition. The characteristics of the vertical ground reaction force values for each volunteer, as seen in Figure 1 and Table I are shown below.

For volunteer A, the anterior peak values increased in the swinging of the large arm swing walking compared to the natural walking condition. For volunteer B, the peak values decreased and the valley values increased with increasing angle of arm swing compared to natural walking. However, compared to volunteers C and D, who showed a similar tendency, the vertical ground reaction force fluctuation due to an increase in the angle of forward arm swing was reduced by a small amount. For volunteer C, the initial peak tended to increase as the arm swing angle became more forward, but decreased in the largely backward arm swing condition. The late peak decreased with increasing arm swing angle. For volunteer D, the value of the early peak increased with increasing backward angle of arm swing, but decreased with increasing forward angle of arm swing.

In all volunteers, the late peak values decreased as the arm swing angle increased both forward and backward. Furthermore, the bimodality of vertical ground reaction force values tended to be lost in the rearward arm swing condition due to the increase in the valley and the late peak values.

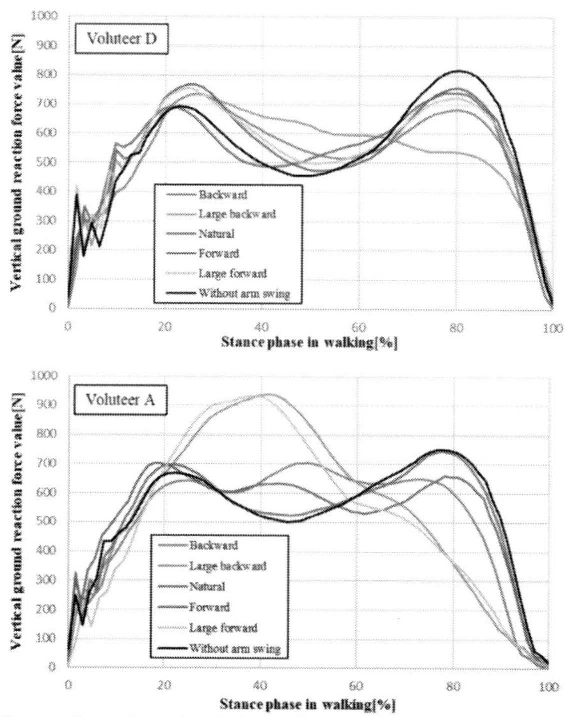

Figure 1. Comparison of vertical ground reaction force values for one trial of each gait condition during the stance phase.

TABLE I. BIMODAL AVERAGED VERTICAL GROUND REACTION FORCE VALUES FOR EACH GAIT CONDITION DURING THE STANCE PHASE. VALUES IN BOXES ARE INITIAL PEAK/VALLEY/LATE PEAK.

ID	Volunteer A	Volunteer B	Volunteer C	Volunteer D
a	687/537/734	634/492/732	712/473/726	719/475/765
b	688/505/747	590/475/750	706/461/760	724/442/807
c	701/639/662	624/509/710	714/508/684	731/522/701
d	926/596/575	614/570/674	709/560/645	788/535/583
e	683/613/670	613/498/721	724/499/692	748/481/758
f	867/612/596	610/525/712	738/520/658	716/521/716

The results of the pelvic lateral tilt angles are shown next. Figure 2 shows the pelvic lateral tilt angles for each randomly selected condition for Volunteer D, who represented Volunteers B~D who showed a similar tendency, and Volunteer A, who showed a different tendency from the other volunteers. Pelvic lateral tilt angle 90 ° and is the pelvic tilt angle horizontal to the ground; When the pelvis is tilted anticlockwise (ACW) as viewed from the direction of travel, the value tends toward the minus side.

Since symmetric movement between the stance and swing phases and similar changes can be observed in the waveforms, Table 2 summarizes the mean values of the troughs and peaks in the early phase and the troughs in the late phase of the stance phase.

The characteristics of each volunteer's transverse pelvic tilt angle seen in Figure 2 and Table 2 are shown below. In the

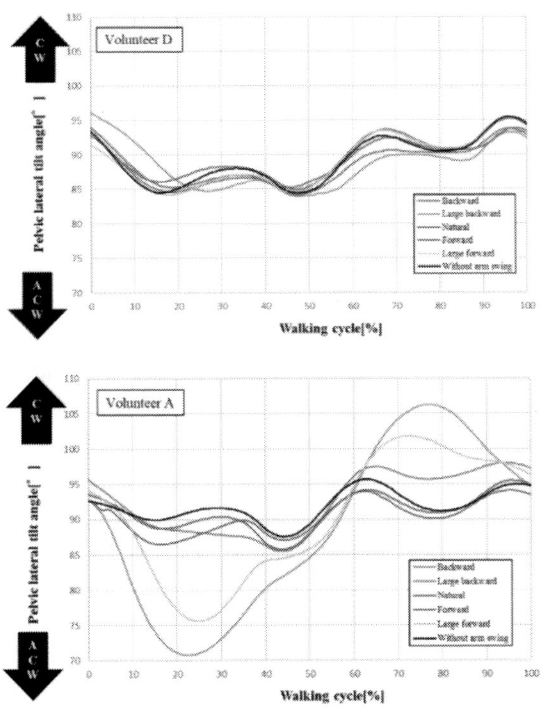

Figure 2. Lateral pelvic tilt angle for one trial of each gait condition in one gait cycle.

TABLE II. MEAN PELVIC LATERAL TILT ANGLE FOR EACH GAIT CONDITION DURING STANCE PHASE. VALUESV IN BOXS ARE EARLY VALLEY/PEAK/LATE VALLEY.

ID	Volunteer A	Volunteer B	Volunteer C	Volunteer D
a	90.2/92.7/87.4	87.1/88.2/82.1	87.0/89.2/84.8	85.1/88.2/84.9
b	88.8/91.1/86.5	87.8/89.1/82.5	86.6/88.5/84.9	84.7/88.6/85.0
c	86.0/87.1/85.2	86.3/87.5/81.6	86.0/87.6/83.8	85.1/86.4/84.3
d	72.7/80.1/80.1	87.2/88.6/80.7	86.3/87.6/82.3	85.2/85.9/83.4
e	87.8/90.1/86.8	86.7/88.1/82.5	87.1/89.0/84.7	84.9/87.5/84.8
f	81.7/84.6/83.9	86.7/88.2/82.3	86.1/87.6/83.0	85.7/88.0/85.6

case of volunteer A, the peak tended to disappear in the large arm swing condition, and the lateral tilt angle also decreased. For volunteer B, the peak values decreased and the valley values increased with increasing angle of arm swing compared to natural walking. For volunteer C, the difference between the early valleys and peaks tended to decrease as the angle of swing of the arm increased, and the late valleys showed a decreasing trend compared to natural walking. For volunteer D, the trends of the early valleys and peaks were similar to those of volunteer C. However, the late valleys decreased with increasing backward angle of arm swing, but increased with increasing forward angle of arm swing.

In all volunteers, the late valleys tended to decrease with increasing angle of arm swing compared to natural walking.

Finally, the vertical ground reaction force values and the pelvic lateral tilt angle under the same conditions are shown in Figure 3.

As shown by the red dashed line in Figure 3, the late peak of vertical ground reaction force and the timing of the decrease in the pelvic lateral tilt angle were identical for all volunteers and all conditions. The data showed that the minimum value and the timing of the pelvic lateral tilt angle were the same in the late peak of the vertical ground reaction force and that the pelvic lateral tilt angle tended to lose its bimodality in the data of ground reaction force values in which bimodality was lost, suggesting that the pelvic lateral tilt angle affected the late peak of vertical ground reaction force values.

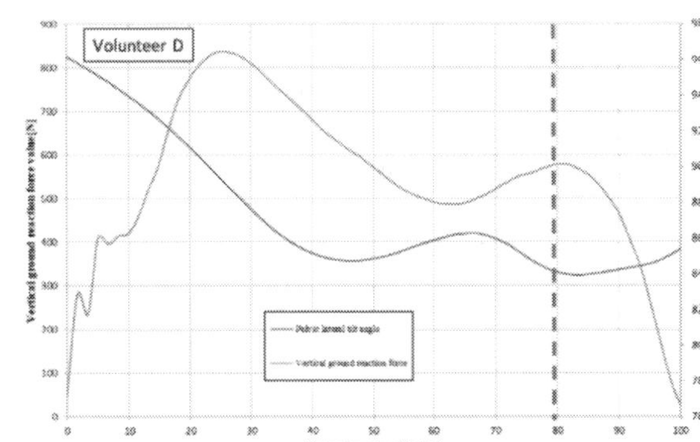

Figure 3. Lateral pelvic tilt angle and vertical ground reaction force values during the stance phase under the same conditions.

IV. CONSIDERATION

The biomechanical relationship between the arm and the bimodal change in the lateral pelvic tilt angle and the vertical ground reaction force is discussed. Regarding the change in the pelvic lateral tilt angle, the shoulder on the side where the arm is forward is tilted downward due to the swing angle of the increase in the backward arm. The pelvis moves upwards as a reaction to this shoulder tilt, which is thought to increase the angle of lateral tilt.

Regarding the relationship between vertical ground reaction force and pelvic adduction, vertical ground reaction force is correlated with acceleration of the body's center of gravity [14]. This suggests that the vertical ground reaction force at the stance terminal phase also decreased because the sacrum, which is considered the center of gravity of the human body, moved up and down due to the lateral pelvic tilt.

V. CONCLUSION

The purpose of this paper is to experimentally clarify how the bimodality of vertical ground reaction force changes with arm swing and to clarify mechanically what kind of coupling caused this change. In the experiment, six conditions of arm swings were set up and the mechanical parameters of each segment and vertical ground reaction force were measured for each gait condition. Experimental analysis showed that the lateral tilt angle of the pelvis changed as the arm rotation angle increased, and the bimodal disappearance of the vertical ground reaction force was observed.

The results of the present paper suggest that the lateral pelvic tilt motion induced by arm swing is involved in the bimodality of vertical ground reaction force. Therefore, it is a future task to clarify the coupling process between the lateral pelvic tilt angle and the vertical ground reaction force. As an approach to solving the problem, mechanical parameters are observed for the segments of the pelvis and lower extremities.

Furthermore, to clarify the coupling between the upper and lower limbs, we will try to approach not only from the joint angles, but also from the muscle activities induced by arm swing. The future research will include the development of an application that provides biofeedback of effective movement indicators to each individual based on the coupled process.

REFERENCES

[1] 2 International Trends in Aging - Cabinet Office, https://www8.cao.go.jp/kourei/whitepaper/w-2022/html/zenbun/s1_1_2.html#:~:text=, Reading Date: 28 July 2023

[2] Ministry of Health, Labour and Welfare, 2021 Summary of Brief Life Tables, https://www.mhlw.go.jp/toukei/saikin/hw/life/life21/index.html, Reading Date: 20 July 2023

[3] Walking Ministry of Health, Labor and Welfare e-Health Nethttps://www.e-healthnet.mhlw.go.jp/information/dictionary/exercise/ys-080.html, Reading Date: 20 July 2023

[4] T. Sato, S. Yamamoto , N. Kawashima, and K. Nakazawa, "Functional role of upper limb motion inhuman bipedal locomotion," (in Japanese) The Japan Society of Mechanical Engineers, No.09-55, Proceedings of the 22nd Bioengineering Conference pp. 163, 2010

[5] S. Inoue and K. Saitou, "Effects of upper limb movement on walking motion during walking- As an aid to fall prevention-," (in Japanese) Japan Society of Physical Education Conference Issue pp.331, 2004

[6] S. T. Eke-okoro, M. Gregoric, and L. E. Larsson, "Alterations in gait resulting from deliberate changes of arm-swing amplitude and phase," Clinical Biomechanics Volume 12 Issues 7-8 pp.516-521, 1997

[7] T. Siragy, C. Mezher, A. Hill and J. Nantel, "Active arm swing and asymmetric walking leads to increased variability in trunk kinematics in young adults," Journal of Biomechanics Volume 99 Article. 109529, 2020

[8] P. Meyns, S. M. Bruijn, and J. Duysens, "The how and why of arm swing during human walking," Gait & Posture Volume 38 pp.555-562, 2013

[9] C. A. Bailey, A. Hill, R. B. Graham, and J. Nantel, "Effects of arm swing amplitude and lower limb asymmetry on motor variability patterns during treadmill gait," Journal of Biomechanics Volume 130 Article. 110855, 2022

[10] Y. Hattori, "Gait Analysis in Clinical Practice," (in Japanese) Physical therapy, Vol. 33, No. 4 pp.207-210, 2006

[11] K. Adachi, M. Okada, S. Kuno, and M. Isizu, "The Effect of Walking Speed and Aging on Ground Antica-taiyo Village Health Promotion Project (14)," (in Japanese) Physical Fitness Science Vol. 48, No. 6 pp.751, 1999

[12] R. Nakamura, S. Hiroshi, and N. Hiroshi, "BasicKinesiology, 6th Edition, Revised," (in Japanese) pp.392, 2003

[13] M. Takami and K. Fukui, "A study of walking of normal volunteers by ground reaction force meter- especially with respect to differences by age and gender –," (in Japanese) Rehabilitation Medicine vol. 24 no. 2 pp.93-101, 1987

[14] Y. Ehara and S. Yamamoto, "Body dynamics beginning analysis of gait and gait Initiation," (in Japanese) ISHIYAKUPUBLISHERS, INC. pp.108, 2002

Identification of Factors Guiding Treatment Decision in Oncology by Rapid Data Insights Using AI and XAI — a Pilot Study on Real-World Data

Holger Ziekow
Faculty of Business Information Systems
Furtwangen University
Furtwangen, Germany
e-mail: zie@hs-furtwangen.de

Norbert Marschner, Dunja Klein
iOMEDICO AG
Medical Department
Freiburg, Germany
e-mail: {norbert.marschner,dunja.klein}@iomedico.com

Benjamin Kasenda
Medical Oncology
University Hospital of Basel
Basel, Switzerland
e-mail: benjamin.kasenda@usb.ch

Nina Haug
iOMEDICO AG
Biostatistics
Freiburg, Germany
e-mail: nina.haug@iomedico.com

Abstract— **Real-world data on the treatment histories of patients in everyday care contain a large amount of latent knowledge which, to date, is almost only made available via publication with a considerable time lag and only in relation to specific issues. Artificial Intelligence (AI) models can capture the knowledge contained in this kind of data and transfer it to new scenarios. We aim to develop an AI-based tool that enables dynamic data exploration and analysis of real-world datasets on medical treatments. The purpose of the tool is to support oncologists in their decision-making process through a system that is trained with prospectively documented real-world data on historical treatment decisions for a large population of patients. It will facilitate research on treatment routines for specific patient populations by providing information on likely therapy choices. Leveraging Explainable AI (XAI) techniques, the reasoning of the analytics system is made transparent to the user. In this paper, we describe and test a system that follows this concept. Specifically, we address the two use cases (a) "therapy selection" and (b) "identification of similar patients". We test respective AI and XAI mechanisms with real-world data. Our analysis provides insights into the potential of the approach of using AI/XAI as supporting analytics system for oncologists as well as on the data requirements.**

Keywords-Explainable AI; Oncology; Medical Information Systems.

I. INTRODUCTION

In this paper, we investigate the potential of using Artificial Intelligence (AI) and Explainable AI (XAI) technologies to support oncologists with a tool for exploring medical registry data. The paper is an updated version of our earlier technical report [1]. In our work, we envision an analytics system that uses AI for providing oncologists with case-specific information and leverages XAI techniques, specifically Shapley (SHAP) values [2], to make its reasoning transparent. The core idea is to learn from historical records of applied treatments and transfer the learned patterns into new settings. These historical records constitute "Real-World Data" (RWD); that is, patients are drawn from a large sample of individuals who received treatment for advanced Colorectal Cancer (CRC) during routine clinical care. Our aim is to leverage AI to make the latent knowledge that rests within RWD accessible to oncologists. Specifically, we address the two use cases of (a) "therapy selection" and (b) "identification of similar patients".

The use case of "therapy selection" is about informing oncologists about likely therapy choices which would have been made by other oncologists for a given (possibly fictious) patient. Here, the AI estimates the probability that an oncologist would prescribe a given therapy to a patient given their characteristics. Using XAI techniques, the underlying reasoning of the algorithm is made transparent. Precisely, the algorithm explains which particular patient features spoke in favor of or against a given therapy choice in a given case. In the use case of "identification of similar patients", the AI model is employed to define a meaningful similarity metric between patients. This metric is based on clinical characteristics available to the treating oncologist.

Within this paper, we present and evaluate solutions for the implementation of both use cases. The evaluation includes quantitative tests as well as qualitative analyses by oncology domain experts. Our main contributions are:

- We present a concept for using AI as supporting analytics system for oncologists
- We analyze the applicability of an AI-based analytics system for estimating probability distributions of therapies, testing various algorithms
- We analyze the dependence of the analytics system's performance on the amount of available training data
- We analyze how XAI can render the reasoning of an AI-based data analytics system transparent to the treating physician

- We present and evaluate an AI-based similarity metric for patient records

The remainder of the paper is structured as follows. We discuss related work in Section II. The rationale and motivation from a medical perspective is given in Section III, along with details of the application scenarios and specifics of the analyzed data sets. In Section IV, we present our AI-based approaches to support decision making for oncologists. This is followed by a description of our experimental setup and experimental results in Section V, and the conclusion of the paper in Section VI.

II. Related Work

Clinical Decision Support Systems (CDSS) constitute an active area of research with applications including diagnostics, prediction of adverse events, and drug control [3], [4]. Existing approaches are often categorized as either knowledge-based or non-knowledge-based. Knowledge-based CDSS build on expert knowledge fed into the system in the form of if-then rules. For example, such systems have been shown to successfully decrease the risk of medication errors in a hospital setting [5]. Non-knowledge-based (or data-driven) approaches, in contrast, leverage real-world data by techniques from statistics and machine learning (ML). For example, in [6], a system based on collaborative filtering for treatment recommendations to psoriasis patients was presented. However, the application of non-knowledge based CDSS remains relatively scarce until today [7]. Challenges faced include lack of available data or low data quality and the black-box behavior of many ML algorithms, limiting trust placed into them by humans.

In the field of cancer therapy, the CDSS Watson for Oncology (WFO) aimed at providing treatment recommendations to oncologists regarding surgical procedures, radiotherapy, and medication. A description of the underlying technology is given in the supplement of [8]. While a meta-analysis found an overall solid agreement of WFO's recommendations with those by human experts [9], this concordance has been shown to vary by country and tumor entity [10]. For these reasons, the WFO service was discontinued in 2022. While we see analytics as the purpose of our tool rather than recommendation, the underlying methodology could also be employed in the framework of a CDSS. Therefore, our work adds to the body of knowledge about data-driven decision support by providing tests on real-world data about CRC treatments and analyzing a specific XAI approach.

ML, XAI and SHAP values have been used in medicine and specifically oncology in previous works. For example, Nohara et al. use a SHAP-based approach to analyze models for predicting the risk of colon cancer [11]. Moncada-Torres et al. address breast cancer survival with machine learning for survival analysis and use SHAP to analyze model predictions [12]. Alabi et al. predict the survival of nasopharyngeal cancer with machine learning and analyze the resulting models with the XAI methods SHAP and LIME [13]. Unlike these works, we do not aim at predicting an outcome, but the therapy. Hence, our XAI analysis reflects factors that impact the therapy selection. Our experiments examine the utility of the approach for this use case. To the best of our knowledge this is the first work presenting such an analysis on real-world cancer data (specifically advanced/metastatic CRC).

Using SHAP for the comparison of data points has been proposed in the context of what has been coined as "supervised clustering" by Lundberg et al. [14]. In the medical domain, Cooper et al. use a clustering method based on that idea to identify subgroups in COVID-19 symptoms [15]. Likewise, our similarity metric is based on this idea of supervised clustering. In this paper, we adapt and test the concept for the use case of finding patient records that are similar in a meaningful way.

III. Medical Background

New treatment options for cancer patients have emerged over the last decades, providing oncologists and patients with an increased number of treatment possibilities. However, with the growing number of options, treatment decision making becomes increasingly complex, thereby challenging medical expertise [16]. What is the best treatment for a patient? Currently, treatment recommendations and guidelines are mainly based on evidence from randomized clinical trials (RCTs) comparing new drugs to standard treatments or placebo. Although RCTs are the best way to compare drugs or treatment strategies, due to their strict in- and exclusion criteria, patients recruited into them are often not representative for those who are intended to receive these treatments in routine clinical care. Consequently, such RCTs can have a high degree of internal validity, but only a low level of external validity [17]. To close this evidence gap, insights drawn from data collected during routine clinical care should also be considered when making treatment decisions. However, it is important that such Real-World Data (RWD) are of high quality to exclude biased inference (e.g., selection or reporting bias). To investigate the potentials of ML in supporting treatment decision making, we set up this project using a high-quality cohort of patients being enrolled in the prospective and multicenter Tumor Registry Colorectal Cancer (TKK) [18].

A. *Aims and Scope*

Treatment decision making relies on many factors, such as patient characteristics (e.g., age and co-morbidities), tumor characteristics, available evidence, patient preferences, and the physician's expertise. The TKK database provides information on the first two aspects: patient characteristics and tumor characteristics. Both are very important factors when it comes to treatment decision making, given that the available evidence is the same and that physician's expertise is usually similar among trained oncologists. Based on this, we have three aims:

- To investigate whether AI can predict — based on given patient and tumor characteristics — what treatment a clinician would have given to a patient.
- To investigate whether XAI methods can render the reasoning of the AI model interpretable.
- To investigate whether AI techniques can be used to define a meaningful similarity metric for patients.

B. Patient Sample

For all experiments outlined below, we used a dataset of patients with advanced/metastatic CRC from the TKK. This is a prospective, multicenter, longitudinal, nation-wide cohort study in Germany which started in 2006. Since then, 269 medical oncologists have recruited more than 4,000 patients with advanced/metastatic disease. This study was reviewed by an ethics committee and is registered at ClinicalTrial.gov (NCT00910819). Eligible patients are 18 or older with histologically confirmed CRC. Patients also received at least one systemic chemo- or targeted therapy (e.g., antibodies) for advanced/metastatic disease. Written informed consent was obtained from all patients. All patients are treated according to physician's choice and are followed for a minimum of 3 years (or until death, loss to follow-up or withdrawal of consent). At the time of enrolment, data on patient and tumor characteristics are documented. From 2008 to 2013, the KRAS-mutation status was collected without further information on the tested/mutated exon(s). Since 2014, data on the extended RAS-testing routine were documented (KRAS-exons 2, 3 and 4 and NRAS-exons 2, 3 and 4), further referred to as (K)RAS and (N)RAS mutation testing, respectively.

For all experiments described herein, we used a sample of 3,563 prospectively enrolled patients with start of the first systemic treatment for their advanced/metastatic CRC. Further details of the TKK have been reported previously [18].

IV. Technical Approach and Concepts

In this section, we provide an overview of our concept for using ML and XAI to support decision making in therapy selection. Figure 1 shows key components and the workflow for their use. We discuss each component and the specific instantiation of our test implementation blow. Further details can be found in [1].

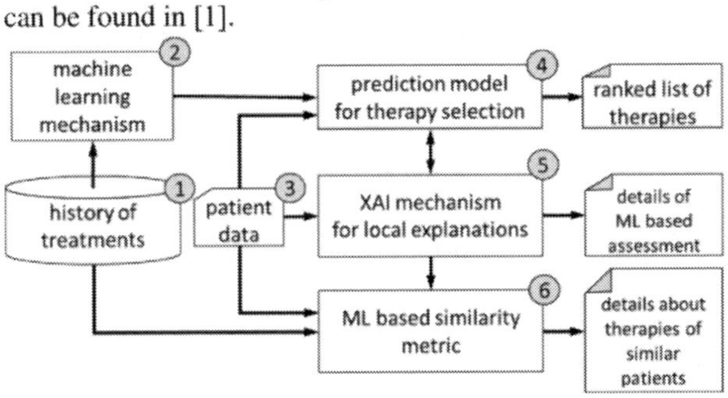

Figure 1. Architecture and key components of the system concept

Component 1 includes patient characteristics such as demographics, medical history, comorbidities, and tumor characteristics. In our study, we used patient data from the TKK registry, alongside with the chosen therapy (see Section III B). In our study, chosen therapies are specified by the therapy backbone (e.g., FOLFOX/CAP+IRI) and the used antibody (e.g., anti-EGFR, anti-VEGF), if applicable. This results in up to $n = 15$ distinct therapy schemes, $n = 12$ when reducing the principle (monotherapy, doublet chemotherapy, or triplet chemotherapy), and $n = 8$ when only indicating if an antibody was given and discarding the antibody details. Our main analysis is carried out with 8 distinct therapies, but we also report results of the other two variants when evaluating the algorithm's performance.

Component 2 refers to an ML algorithm that learns to predict which therapy is chosen for a patient based on information about patient and disease characteristics at the beginning of treatment. It thereby aims at mimicking the decision made by oncologists. In principle, any supervised learning mechanism could be used for this task. In our main implementation we used XGBoost [19], which generally shows good performance on tabular data [20]. We compare its classification performance to other well-established methods and a baseline in our experiments.

Component 3 refers to data of a specific patient for whom the system should support therapy selection. This information contains the subset of features that are available in the history of treatments (component 1), that is relevant for the specific prediction model.

Component 4 is the prediction model resulting from training an ML algorithm on historical treatment decisions. The model is used to estimate, for new instances, the probability distribution of therapies given a set of patient characteristics.

Component 5 provides local explanations for the prediction of the ML model for a given patient. Such mechanisms give insights on how important a given feature was for the decision for a given instance (e.g., how a certain mutation impacted therapy prediction for that patient). This contrasts with global feature importance that assesses the general importance of a feature (e.g., average importance across many cases). In our implementation, we use the SHAP library to compute Shapley values [2], reflecting the fair contribution of each feature to the outcome prediction [21]. Here we take two inputs: (a) the prediction model and (b) the patient data for which we explain the prediction. The output is the Shapley value (case-specific importance) of each feature in the given patient record. Oncologists can use this information to reason about the model's decision for and against different therapies.

Component 6 identifies patients that are similar to a reference patient (see aim #3). This enables oncologists to inform themselves about past treatment routines applied to similar patients. The key challenge is to find a suitable definition of similarity. Here, we build upon the idea of "supervised clustering" as presented in [14]. The concept is to define similarity via local feature importance instead of raw feature values. In our case, the importance stems from what the prediction model has learned about case-specific therapy selection. With this concept, similarity focuses on features that the model finds relevant in a given case. This is in contrast to a similarity metric that factors in the features for all patients in the same way, regardless of the specifics of their case. For our implementation, we concatenate vectors of Shapley values for each feature and each target class. The intuition is that similar cases have similar importance for the same features in therapy prediction. Precisely, we represent each patient by a vector $v = (v_1, v_2, \dots, v_m)$ with $m = n \cdot p$. Here, p is the number of patient features and n is the number of target classes (therapies). The entry $v_{(k-1)p+j}$ is the Shapley value of feature j in the prediction of therapy class k, where $1 \leq k \leq n$. Similarity between two patients is then defined in terms of the Manhattan distance of their vector representations. In our main analysis, we have $n = 8$ and $p = 116$.

V. EXPERIMENTS

Our experiments are designed to test feasibility of using AI to aid therapy selection for patients with advanced/metastatic CRC. Specifically, we address four questions: (1) What is the quality of AI-based therapy selection, (2) do local explanations with Shapley values render the AI algorithms' selection interpretable, (3) is the AI-based similarity metric meaningful and (4) how does data availability impact the performance of the AI-based therapy selection? We describe the corresponding experiments below. We use accuracy and f_1-score as metrics for the predictive performance of the algorithm. All analyses were done in python using the libraries xgboost (v1.5.0) [19], scikit-learn (v1.0.2), scikit-optimize (v0.9.0) and shap (v0.40.0).

A. Experiment 1: Quality of Predictions

Assessing the quality of the predictions made by the AI algorithm yields a conceptual challenge since, in general, the best therapy for a given patient is unknown. We therefore here resort to comparing the AI's predictions with the therapy decisions made by humans. However, even human experts may disagree regarding the optimal therapy for a specific case. This limits the quality that we can expect to observe but provides us with an indication about the quality of AI-based predictions.

1) Experimental Setup

In the experiments, we used records of 3,586 individual CRC patients. After removing implausible records, we were left with 3,563 patients. We extracted 67 variables containing information about patient's health status at the beginning of their first palliative therapy. We used one-hot encoding for non-binary categorical variables if they had less than 10 possible values and label encoding otherwise. Ordinal variables were encoded numerically. This gave us $p = 116$ features as predictors for our ML model. As label for model training, we used the chosen therapies, as described in Section IV, *component 1*. This resulted in 8 different first-line therapies within the data set. From this data set we selected a stratified random sample of 60% of the records as training set and held out the rest for testing. We fitted a classifier using XGBoost with balanced class weighting. Predictive performance was measured in terms of macro-averaged f_1-score (the harmonic mean of precision and recall). We optimized hyperparameters of the classifier with respect to this metric by Bayesian hyperparameter search using the class BayesSearchCV from scikit-optimize. As a benchmark, we trained several alternative ML algorithms, where we optimized hyperparameters using the same method. The tested algorithms were a Random Forest (RF), decision tree, logistic regression with L^2 regularization, linear Support Vector Classifier (SVC) and a dummy classifier always predicting the most frequent class. For logistic regression and SVC, we binned continuous variables by using the bin edges [0, 50, 60, 70, 80, 100] for age, [0, 10, 50, 100, 200, 500, 2000] for the number of weeks since primary diagnosis, [0, 18.5, 25, 30, 100] for Body Mass Index (BMI) and [0, 5, 30] for the Charlson Comorbidity Index. The bins were chosen to allow sufficient numbers of examples in each category or to reflect a common categorization (in the case of BMI) [22]. We used one-hot encoding for all categorical variables thereafter, combining missing or unknown values into separate categories. This led us to $p = 158$ variables.

2) Evaluation

Figures 2 and 3 show the confusion matrix and ROC curves for the classifier's predictive performance on the test set, respectively. Macro-averaged f_1-scores for the three different levels of granularity of the therapy classes are reported in Table I. There, we also report performances of the benchmark methods.

According to Figure 2, predictive performance increases with the number of examples for a given class. For the most frequent class (doublet therapy with antibody), we observe fair agreement between the model's prediction and the actual given treatment and thus therapy choice of the treating physician. Rare therapies, on the other hand, are rarely predicted, therefore yielding poor evaluation results. This is expected, since therapy selection is influenced by personal opinions and preferences of the corresponding physician and patient. This means that different physicians would often select different therapy strategies for the same individual. Moreover, the training set includes a low — and possibly insufficient — number of examples for rare therapies. However, it is encouraging that, according to Figure 3, AUC values of the ROC curves are high for some of the less frequent therapy classes. RF performs similar to XGBoost, and these two methods clearly outperform the other algorithms. This suggests the presence of relevant interactions between variables which cannot be captured by linear models. One can expect improved results with bigger training sets. We investigate this effect in our fourth experiment.

B. Experiment 2: Insights with global and local feature importance measures

Here, we aim at testing the benefits of using local feature importance to support the therapy decision of oncologists. This is qualitative by nature and aims at providing insights into the use of XAI in the targeted use case. We computed and visualized Shapley values for therapy predictions. This comprises global Shapley values (i.e., local Shapley values averaged over the entire test set) and local Shapley values for individual predictions. The visualizations are then analyzed by domain experts regarding validity from a medical perspective. We provide a sample result and the corresponding medical analysis below. To protect privacy of patients in the displayed figures, Gaussian noise was added to the features Age at start of 1st-line, Date of inclusion, Weeks since primary diagnosis, and BMI. Note that the noise was added to the entire data set after training the model, but before computing the Shapley values.

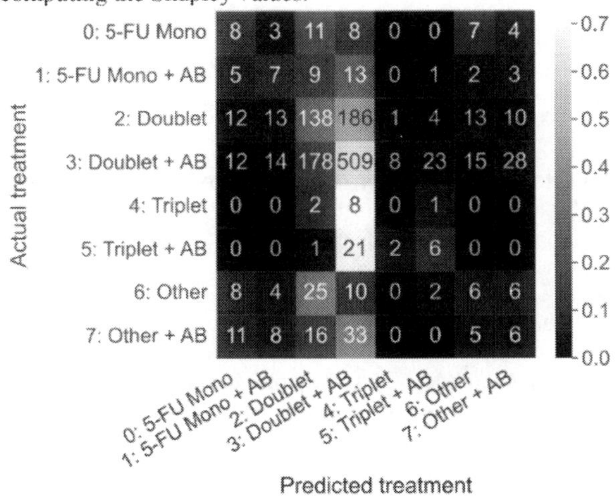

Figure 2. Confusion matrix for 8 distinct therapy classes. The cell in row j and column k is colored according to the fraction of patients who were predicted to receive therapy k among those patients who actually received therapy j.

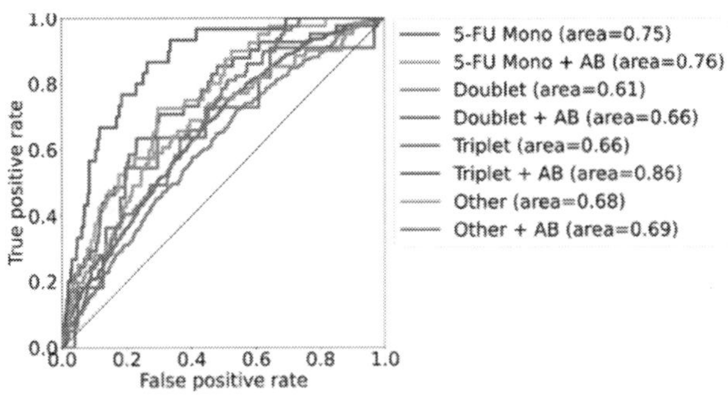

Figure 3. ROC curves for the classifier's predictive performance (case of 8 distinct therapy classes).

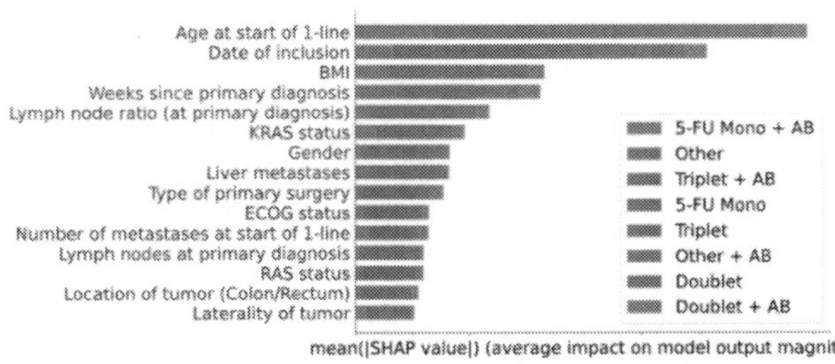

Figure 4. The 15 most important features for therapy prediction, with importance measured in terms of their global Shapley value. Color coding shows the contribution of the different therapy classes.

Figure 4 shows the 15 most important features used for therapy prediction. Here, we obtain a global measure of feature importance by averaging the magnitude of the Shapley value of each feature and therapy over all patients in the test

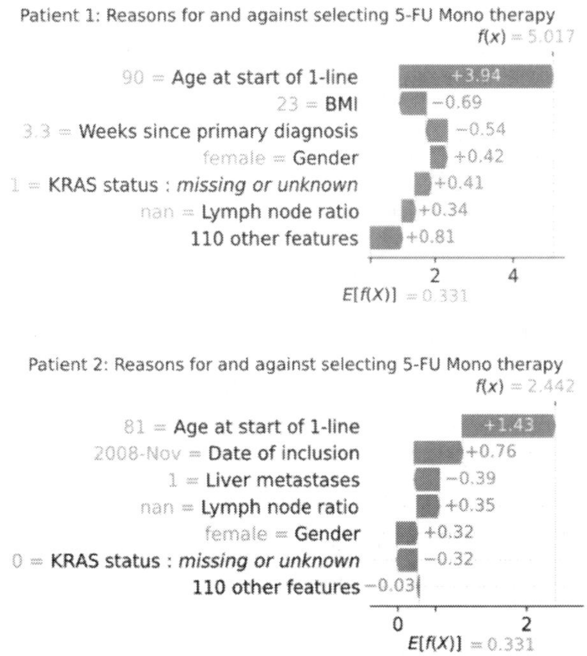

Figure 5. Shapley values for and against 5-FU monotherapy for two representative patients where the algorithm correctly predicted 5-FU monotherapy.

set. The importance of a feature is defined as the class specific global Shapley value, summed over all therapy classes. The length of the horizontal bars represents the importance of the given feature.

In Figure 5, we show representative Shapley values for two patients where the algorithm correctly predicted therapy 5-FU monotherapy (that is, intravenous 5-FU without an antibody) out of 8 possible choices. For privacy reasons, data shown have been overlaid with noise. These examples represent interesting cases where a less common therapy was chosen. Such cases are well suited to check if the special reasons for using such a therapy are well reflected in the Shapley values. Both patients in Figure 5 are over 80 years when starting therapy. Age is known to be a very important factor in clinical decision-making because it strongly correlates with frailty and increased risk of treatment-related side effects. For patient 1, BMI, which was within the normal range, was a factor rather speaking against choice of 5-FU monotherapy, although the effect was not very strong. BMI is also a surrogate for morbidity; in the context of CRC, low BMI can be associated with frailty and is a sign of malnutrition and disease activity. Thus the "normal" BMI may have been considered by the model as a factor allowing more intense treatment than 5-FU monotherapy.

Figures 6 and 7 show Shapley values for two different representative patients for which the algorithm predicted 5-FU monotherapy when actually doublet chemotherapy was applied. Such cases provide insights into potential causes for

TABLE I. PREDICTIVE PERFORMANCE OF THE COMPARED CLASSIFIERS, AS MEASURED BY THE MACRO-AVERAGED f_1-SCORES FOR THE THREE DIFFERENT LEVELS OF AGGREGATION OF THERAPY CLASSES.

Number of distinct therapies	**8**	**12**	**15**
XGBoost	0.21	0.19	0.15
Random Forest	0.23	0.20	0.17
Logistic Regression	0.17	0.16	0.13
Linear Support Vector Classifier	0.17	0.16	0.15
Decision Tree	0.19	0.14	0.14
Dummy Classifier	0.09	0.05	0.02

divergence between the AI-based prediction and the treatment decision. The figures show the most important Shapley values for the prediction of 5-FU monotherapy and doublet chemotherapy, respectively. For patient 1, similar to the true positives in Figure 5, increased age (85 years) was speaking for 5-FU monotherapy. We can see in Figure 7 that this factor is reversed, speaking against the treatment with doublet chemotherapy. In patient 2, missing grading status and time since primary diagnosis were factors favoring doublet chemotherapy. In addition, presence of anemia — which is also considered a surrogate for morbidity — was speaking against doublet chemotherapy. As outlined above, age is an important factor for treatment-decision making, but chronological age does not necessarily mirror the frailty status of a patient (fit elderly patients). Although we have information on clinical performance status for most patients in our dataset, other important factors also driving treatment decisions are not captured (e.g., patient preference). For

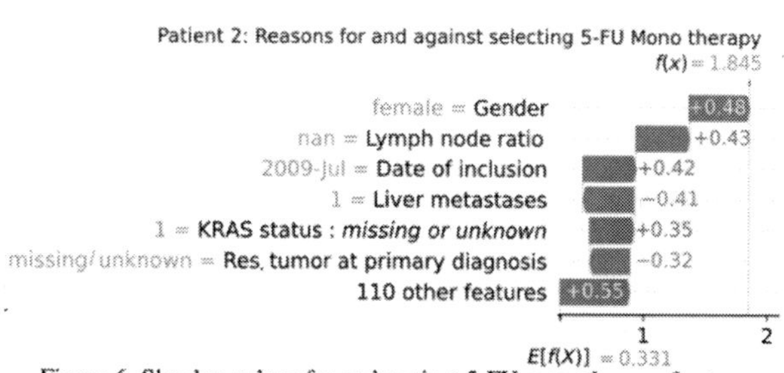

Figure 6. Shapley values for and against 5-FU monotherapy for two different representative patients for which the algorithm predicted 5-FU monotherapy when actually doublet chemotherapy was applied.

instance, some patients may opt for a more intense treatment despite higher risk for side effects. Such factors outside our database might have driven the treatment decision. Interestingly, one would have assumed co-morbidities and clinical performance status to have more weight in the decision, but their effect are rather modest in either direction.

C. *Experiment 3: Benefits of AI-based similarity metric*

Here, we analyze the benefits of using the proposed AI-based similarity metric. The goal is to show that this metric helps to identify patients that are similar in a meaningful way. A direct way to evaluate this would be to ask domain experts to assess the results. Here we take an indirect approach. That is, we use our similarity metric as input for a K-Nearest-Neighbor (KNN) classifier for therapy prediction and compare classification results against a baseline metric. We argue that the KNN classifier yields better prediction results if the underlying metric is more meaningful from a medical perspective.

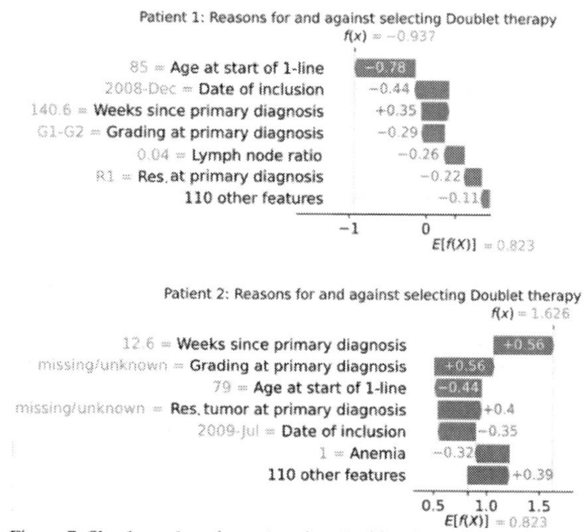

Figure 7. Shapley values for and against doublet chemotherapy for two representative patients where the algorithm predicted 5-FU monotherapy when actually doublet chemotherapy was applied.

1) *Experimental Setup*

For the distance between two patients, we use the metric based on Shapley values as described above. We use the same partitioning of the data into training and test set as in experiment 1 and fit a KNN-classifier with number of neighbors $k = 5$ and inverse distance weighting to the training data. Implementation is done with the KNeighborsClassifier from scikit-learn. For our baseline metric, we represent each patient by the vector $\mathbf{w} = (w_1, w_2, \ldots, w_p)$ of their features, with the same preprocessing as for logistic regression and SVC in Experiment 1. As for the Shapley value-based metric, similarity of two patients is then defined in terms of the Manhattan distance of their vector representations and a KNN-classifier with the same parameters is fitted to the training data. We compare performance of the two classifiers using the same metrics as in Experiment 1.

2) *Evaluation*

The experiments show improvements of the prediction quality when using KNN with the Shapley value-based distance metric, compared to a naïve baseline-distance metric (Table II). Although improvements are small, results indicate that the Shapley-based distance metric may provide a meaningful similarity measure. Note that this experiment evaluates the desired effect only indirectly and classification is not the aim of the addressed use case. For many instances, a less elaborate metric may find less similar patients but lead to the same therapy prediction. In such cases, we would observe no benefits. However, our approach aims at identifying patients that are similar in a meaningful way, so that they can serve as reference cases. Here, better similarity is beneficial even if the recorded therapies are the same. Since the tests with a KNN classifier can only reveal benefits for certain cases, we find the observed improvement encouraging. An analysis with human experts who directly assess the usefulness of the similarity metric may further clarify the benefits of the approach.

TABLE II. COMPARISON OF THE PERFORMANCE OF KNN CLASSIFIERS BASED ON THE SHAP-BASED METRIC AND THE BASELINE METRIC

Score type	Classifier	Number of distinct therapies		
		8	12	15
f_1 (macro average)	KNN (Shapley)	0.18	0.16	0.15
	KNN (Baseline)	0.16	0.13	0.12
f_1 (weighted average)	KNN (Shapley)	0.49	0.43	0.23
	KNN (Baseline)	0.49	0.38	0.22
Accuracy	KNN (Shapley)	0.54	0.46	0.25
	KNN (Baseline)	0.55	0.42	0.24

D. *Experiment 4: Impact of data availability*

The amount of training data impacts the performance of ML models, but strength of this impact varies from case to case. Here, we analyze this effect for the case of CRC therapy prediction.

1) *Experimental Setup*

For our experiments we have in total 3,563 patient records available (see Experiment 1 – Experimental setup). To investigate the effect of training size, we trained multiple models on differently sized subsets of the data. Specifically, we set aside a 40% stratified sample for testing and used the rest as source for training data. From the training data, we iteratively took 90% stratified subsets to train models (thereby iteratively reducing training set size). We then computed performance measures (f_1-score for one-versus-all) on the initially held out test sets for each model and therapy class. The process was repeated 10 times with different random seeds.

2) *Evaluation*

Figure 8 shows the results of the experiment about the impact of the training data size. From visual inspection of these plots, it appears that the learning curves for the prediction of two

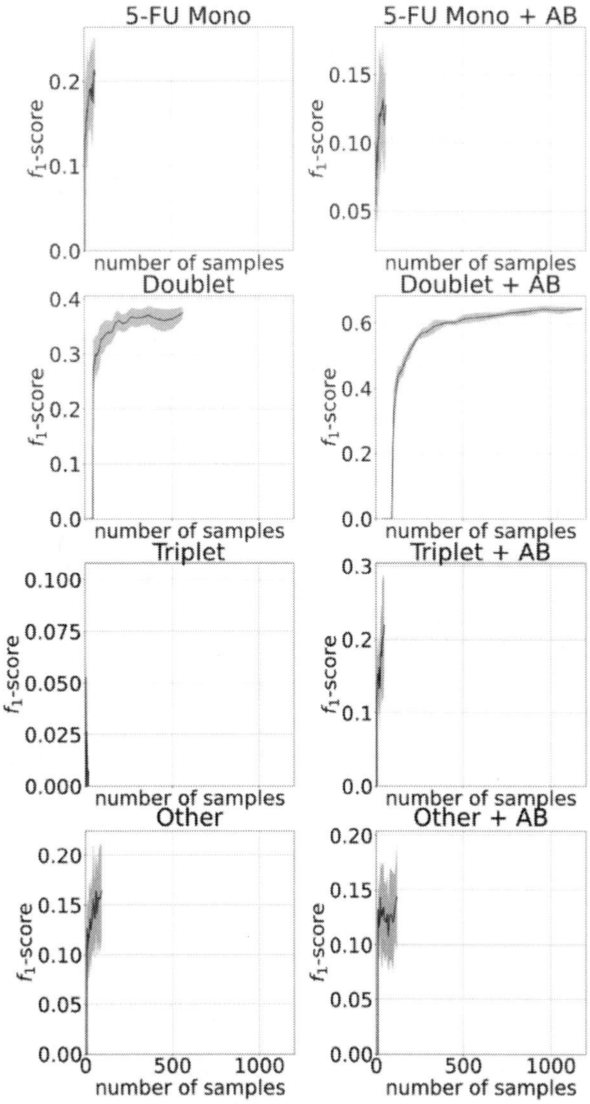

Figure 8. Impact of the number of training samples of a given therapy class on the model's performance, measured in terms of f_1-score for one-vs-all. Dark blue is the mean value and light blue is the standard deviation.

therapies reach a saturation point at about 500 training instances. For all other therapies, we have less than 500 instances in the training set and do not observe saturation. Note that for several therapies, the number of training examples is rather small, resulting in poor prediction performance of the model for those therapies. However, the shape of the curve suggests that improvements with larger training sets may be possible. We draw two main conclusions from this analysis. One is that performance results for therapy prediction (observed in Experiment 1) would likely improve with additional training data. The other conclusion is that about 500 training instances per therapy may be sufficient for the chosen ML setup.

VI. CONCLUSION

In this paper, we analyzed the potential of using AI to build an information tool that enables dynamic data exploration and analysis of RWD. Specifically, we analyzed the two use cases "therapy selection" and "identification of similar patients." Both objectives aim to provide a second view built on the large amount of RWD and thus make this broad knowledge accessible to individual oncologists. For these use cases, we proposed a system setup using supervised learning and XAI techniques.

We have shown applicability of the concept and obtained insights on the required amount of training data, but additional work should be done to assess viability of our approach. While we have demonstrated superiority of the AI-based approach against baseline methods, our experiments show a certain degree of disagreement between predicted and chosen therapies. However, disagreement is expected if different human experts are asked to give a second opinion. Quantifying the level of human disagreement and comparing this to the AI-based results is subject to future work.

Our approach has limitations which should be addressed in future work. One challenge is posed by the fact that the AI algorithm learns therapy selection from prospectively recorded past records. However, the therapy landscape in oncology develops quickly, causing concept drift; that is, historic decisions learned by the algorithm may have better alternatives by now. Also, best practice about therapy decision may change over time and change the probability of selecting a therapy for a given patient. This concept drift makes the algorithm prone to the cold-start problem of AI-based recommender systems. Solutions to this problem could involve non-uniform weighting of observations based on treatment date or incorporation of expert knowledge.

It is important to stress that therapy outcomes of patients such as overall survival, progression-free survival, and quality of life, which are also documented in the TKK database, were not considered. This means that the information tool may reproduce and even reinforce suboptimal, yet common practice in treatment routine.

Furthermore, feature selection remains subject to future work. Due to the high number of features and therapy classes, the Shapley value-based similarity measure may be subject to the curse of dimensionality. Incorporation of feature reduction techniques may therefore lead to better results. We also note that, while Shapley values are a measure for the impact of a given feature on a model's prediction for a given subject, they do not imply causation.

We believe that the investigated concepts have great potential to support information processes in cancer care using dynamic data exploration and analysis of real-world datasets. Our findings show promising results that call for further analysis and development of the outlined ideas. Beyond expanding on these ideas and addressing the discussed limitations, we plan expansion to further use cases in the future.

REFERENCES

[1] H. Ziekow, N. Marschner, D. Klein, B. Kasenda, and N. Haug, "Technical report: Identification of factors guiding treatment decision in oncology by rapid data insights using AI and xAI - a pilot study on real-world data." 2022. Accessed: Oct. 04, 2023. [Online]. Available: https://opus.hs-furtwangen.de/frontdoor/index/index/docId/8454

[2] S. M. Lundberg and S.-I. Lee, "A unified approach to interpreting model predictions," in *Proceedings of the 31st International Conference on Neural Information Processing Systems*, in NIPS'17. Red Hook, NY, USA: Curran Associates Inc., Dec. 2017, pp. 4768–4777.

[3] R. T. Sutton, D. Pincock, D. C. Baumgart, D. C. Sadowski, R. N. Fedorak, and K. I. Kroeker, "An overview of clinical decision support systems: benefits, risks, and strategies for success," *npj Digit. Med.*, vol. 3, no. 1, p. 17, Dec. 2020, doi: 10.1038/s41746-020-0221-y.

[4] T. N. T. Tran, A. Felfernig, C. Trattner, and A. Holzinger, "Recommender systems in the healthcare domain: state-of-the-art and research issues," *J Intell Inf Syst*, vol. 57, no. 1, pp. 171–201, Aug. 2021, doi: 10.1007/s10844-020-00633-6.

[5] C. D. Mahoney, C. M. Berard-Collins, R. Coleman, J. F. Amaral, and C. M. Cotter, "Effects of an integrated clinical information system on medication safety in a multi-hospital setting," *American Journal of Health-System Pharmacy*, vol. 64, no. 18, pp. 1969–1977, Sep. 2007, doi: 10.2146/ajhp060617.

[6] F. Gräßer *et al.*, "Therapy Decision Support Based on Recommender System Methods," *Journal of Healthcare Engineering*, vol. 2017, pp. 1–11, 2017, doi: 10.1155/2017/8659460.

[7] F. Gräßer, F. Tesch, J. Schmitt, S. Abraham, H. Malberg, and S. Zaunseder, "A pharmaceutical therapy recommender system enabling shared decision-making," *User Model User-Adap Inter*, pp. 1019–1062, Aug. 2021, doi: 10.1007/s11257-021-09298-4.

[8] S. P. Somashekhar, "Watson for Oncology and breast cancer treatment recommendations: agreement with an expert multidisciplinary tumor board," *Annals of Oncology*, vol. 29, no. 2, pp. 418–423, 2018, doi: https://doi.org/10.1093/annonc/mdx781.

[9] Z. Jie, Z. Zhiying, and L. Li, "A meta-analysis of Watson for Oncology in clinical application," *Sci Rep*, vol. 11, no. 1, Art. no. 5792, Mar. 2021, doi: 10.1038/s41598-021-84973-5.

[10] N. Zhou *et al.*, "Concordance Study Between IBM Watson for Oncology and Clinical Practice for Patients with Cancer in China," *The Oncologist*, vol. 24, no. 6, pp. 812–819, Jun. 2019, doi: 10.1634/theoncologist.2018-0255.

[11] Y. Nohara, K. Matsumoto, H. Soejima, and N. Nakashima, "Explanation of Machine Learning Models Using Improved Shapley Additive Explanation," in *Proceedings of the 10th ACM International Conference on Bioinformatics, Computational Biology and Health Informatics*, in BCB '19. New York, NY, USA: Association for Computing Machinery, Sep. 2019, p. 546. doi: 10.1145/3307339.3343255.

[12] A. Moncada-Torres, M. C. Van Maaren, M. P. Hendriks, S. Siesling, and G. Geleijnse, "Explainable machine learning can outperform Cox regression predictions and provide insights in breast cancer survival," *Sci Rep*, vol. 11, no. 1, Art. no. 6968, Mar. 2021, doi: 10.1038/s41598-021-86327-7.

[13] R. O. Alabi, M. Elmusrati, I. Leivo, A. Almangush, and A. A. Mäkitie, "Machine learning explainability in nasopharyngeal

cancer survival using LIME and SHAP," *Sci Rep*, vol. 13, no. 1, Art. no. 8984, Jun. 2023, doi: 10.1038/s41598-023-35795-0.

[14] S. M. Lundberg, G. G. Erion, and S.-I. Lee, "Consistent Individualized Feature Attribution for Tree Ensembles," Feb. 2018, Accessed: Oct. 04, 2023. [Online]. Available: http://arxiv.org/abs/1802.03888

[15] A. Cooper, O. Doyle, and A. Bourke, "Supervised Clustering for Subgroup Discovery: An Application to COVID-19 Symptomatology," in *Machine Learning and Principles and Practice of Knowledge Discovery in Databases*, M. Kamp, I. Koprinska, A. Bibal, T. Bouadi, B. Frénay, L. Galárraga, J. Oramas, L. Adilova, G. Graça, et al., Eds., in Communications in Computer and Information Science. Cham: Springer International Publishing, 2021, pp. 408–422. doi: 10.1007/978-3-030-93733-1_29.

[16] P. Bossaerts and C. Murawski, "Computational Complexity and Human Decision-Making," *Trends in Cognitive Sciences*, vol. 21, no. 12, pp. 917–929, Dec. 2017, doi: 10.1016/j.tics.2017.09.005.

[17] S. Khozin, G. M. Blumenthal, and R. Pazdur, "Real-world Data for Clinical Evidence Generation in Oncology," *J Natl Cancer Inst*, vol. 109, no. 11, pp. 1–5, Nov. 2017, doi: 10.1093/jnci/djx187.

[18] N. Marschner *et al.*, "Oxaliplatin-based first-line chemotherapy is associated with improved overall survival compared to first-line treatment with irinotecan-based chemotherapy in patients with metastatic colorectal cancer - Results from a prospective cohort study," *Clin Epidemiol*, vol. 7, pp. 295–303, 2015, doi: 10.2147/CLEP.S73857.

[19] T. Chen and C. Guestrin, "XGBoost: A Scalable Tree Boosting System," in *Proceedings of the 22nd ACM SIGKDD International Conference on Knowledge Discovery and Data Mining*, San Francisco California USA: ACM, Aug. 2016, pp. 785–794. doi: 10.1145/2939672.2939785.

[20] R. Shwartz-Ziv and A. Armon, "Tabular data: Deep learning is not all you need," *Information Fusion*, vol. 81, pp. 84–90, May 2022, doi: 10.1016/j.inffus.2021.11.011.

[21] C. Molnar, *Interpretable Machine Learning*. Accessed: Oct. 04, 2023. [Online]. Available: https://christophm.github.io/interpretable-ml-book/

[22] World Health Organisation, "The SuRF report 2: surveillance of chronic disease risk factors." Accessed: Oct. 04, 2023. [Online]. Available: https://iris.who.int/bitstream/handle/10665/43190/9241593024_eng.pdf

Attempt for Estimation of Vertical Ground Reaction Force by Deep Learning with Time Factor from 2D Walking Images

Takeshi Mochizuki
Kochi University of Technology
Tosayamada, Kami, Kochi, 782-8502, Japan
e-mail: mochizuki.takeshi0094@gmail.com

Kyoko Shibata
Kochi University of Technology
Tosayamada, Kami, Kochi, 782-8502, Japan
e-mail: shibata.kyoko@kochi-tech.ac.jp

Abstract: **Ground reaction force data are useful for evaluating gait stability, but only specialized institutions can measure it because installed force plates are often used to measure it with high accuracy. Therefore, this report proposes an easy method for estimating ground reaction forces using images captured by a widely available device. In a previous report, we created an algorithm to estimate the ground reaction force from images using 2D Convolutional Neural Network (CNN), one of the deep learning. The results showed that if a deep learning model is created in advance, the estimation of vertical ground reaction forces can have an 8% to 14% error to body weight. To further improve accuracy, this report creates training data that include the time factor and performs vertical ground reaction force estimation by 3D CNN. The training data used in this report, the voxel data were created using images at the time of estimation and images prior to that time to incorporate the time factor. The results, estimation of ground reaction force resulted in a 15% error relative to body weight and did not improve accuracy. Since overlearning occurred in all deep learning models, we suppose that accuracy was not improved due to insufficient training data or bias.**

Keywords- Gait Analysis; Ground Reaction Force; Estimation; 3D CNN; Single Camera.

I. INTRODUCTION

Gait exercise is important for maintaining and improving health. However, a gait style that places the load on one leg only or that tends to place the load on the ground is less effective. For good health, it is necessary to keep in mind that gait should be stable on both sides of the body on a daily basis. In clinical practice, gait stability is determined from the time history of the ground reaction force waveforms for each leg during walking, and physicians and physical therapists with expertise in this area provide guidance on gait improvement based on the waveforms. Therefore, we believe that it is possible to diagnose gait stability from the ground reaction force waveform, and if individuals can easily and at any time know the ground reaction force waveform while walking, they will be aware of the need to improve their gait, which will contribute to extending their healthy life span.

The method used to accurately measure ground reaction force waveforms consists of installed force plates. However, installed force plates are expensive, are only available in specialized facilities and cannot be installed at the individual level. This means that ground reaction force values cannot be obtained on a daily basis. In addition, walking movement becomes more deliberate because one must step on an installed force plate. This occurs the problem that normal gait cannot be measured [1].

As an alternative to an installed force plate, this research group has proposed a method to derive the ground reaction force from the acceleration obtained by a wearable inertial sensor using the balance between inertial force and ground reaction force, as reported by Isshiki et al. [2]. This method estimates the combined ground reaction force of the left and right legs, making it difficult to use in clinical settings where ground reaction force for each leg is desired. In addition, the method does not consider individual differences because it uses the mass of each body segment and the position of the center of gravity of each body segment calculated from statistical values as parameters used to derive the ground reaction force. To address this problem, Liu et al. [3] used individual kinematic data and deep learning to estimate ground reaction forces without using statistical values. They estimated ground reaction forces using angular data obtained from optical motion capture and deep learning and showed that they can be estimated at 2% to 8% error to body weight for stair walking. Since personal kinematics data is used, individual differences can be considered. However, because it uses optical motion capture, it cannot be used in everyday life. In contrast, Sakamoto et al. [4] reported an example of estimating ground reaction forces using only data obtained by wearable sensors and deep learning, without using kinematics. They estimated ground reaction forces in a standing static posture and reported that it can be estimated with an average estimation error of 7.6%. However, since the input data used were myopotential, acceleration, and angular acceleration acquired by wearable sensors, the system was not easy to wear and was not simple to use.

On the other hand, Yagi et al. [5] is an example of an attempt to analyze gait from videos that can be easily captured, although it is not an estimation of ground reaction forces. Gait analysis was performed using OpenPose [6], which detects skeletal information from videos and images, and was able to obtain stride length and walking speed from videos captured with an RGB camera. Using only a camera without a wearable sensor for sensing, gait analysis is achieved with fewer burdens on the user. However, as Yagi et al. also mentioned, the method using OpenPose causes errors in the skeletal information acquired from OpenPose, which also causes errors in the estimation results.

To address these issues, this study aims to establish an algorithm to estimate triaxial ground reaction force waveforms for each left and right legs using only cameras that are easy to sense. The proposed algorithm does not use statistics-based kinetic theory, dedicated software to detect skeletal information from videos and images, or other unfamiliar sensor systems, such as wearable sensors, when estimating ground reaction forces. Only videos and images will be used for ground reaction force estimation to eliminate the physical burden on the user during sensing.

As a methodology to achieve this, we have proposed a ground reaction force estimation method using a Convolutional Neural Network (CNN), which is a type of deep learning that excels in image classification, in our previous report [7]. The system creates in advance a deep learning model capable of estimating triaxial ground reaction forces in natural and abnormal walking on level ground using CNN from walking images captured by an RGB camera, so that the user only needs to capture walking images to perform the estimation.

In the previous report [7], the estimation of the ground reaction force from the load response phase to the front swing phase was performed using only walking images obtained from an RGB camera for detection. Accuracy was verified using cross-validation for five volunteers. The results showed that in a laboratory environment, vertical ground reaction forces can be estimated with an error of approximately 8% to 14% error to body weight and an average Pearson's correlation coefficient of 0.80. However, Dongwei Li et al. [3] and Sakamoto et al. [4], mentioned above, estimated it at 2 to 8% error to body weight, even though operating conditions were different. Therefore, in this report, we consider improving the accuracy of the proposed method by targeting 5%, which is a similar level of accuracy of estimation. In the previous report, we improved the estimation accuracy by converting color images to black-and-white images and reducing the image resolution to eliminate the influence of clothing color when creating training data. To further improve the accuracy of estimation, the training data created in the previous report does not include time factor, even though gait is a continuous motion. In this report, we consider learning time factor as well. As a first step of verification to improve accuracy, voxel data containing time factor are used as training data. In this report, vertical ground reaction forces are used.

In the next section, we describe how to create training data that includes time factor and how to create a deep learning model devised in this report. Section III presents the results of the vertical ground reaction forces estimated by the deep learning model created, Section IV discusses the reasons for the lack of improvement in accuracy, and Section V conclusions close the report.

II. METHOD

A. Deep Learning Model

When the user uses the system, he/she simply takes a walking video and inputs it to the system without prior preparation. To achieve this, a deep learning model must be created in advance. A deep learning model is a learning model that outputs ground reaction force values normalized by body weight when voxel data created from a walking image are input. The structure of the 3D CNN in the deep learning model consists of an input layer, followed by two convolution layers, a pooling layer, and a dropout layer to prevent overlearning. The process was repeated from the convolution layer to the dropout layer. After smoothing, it was passed through the fully connected layer one layer at a time, and then the output layer. The convolution layer uses the Relu function as the activation function, and the all-coupled layer uses the Softmax function.

B. Experimental Methods

An experiment was conducted to obtain data to be used for training and validation. The same experimental design as previously reported [7] was used to see the difference in accuracy of the training data generation method. In the experiment, 1 force plate unit (manufactured by Tec Gihan Co., Ltd., TF-6090-C 1 unit) was used for training data for deep learning models and an iPad Pro as a camera were used. Five healthy male volunteers (age 22±1, height 1.73±0.05 [m], weight 61±13 [kg]) participated in the experiment. A 10 step walk path was prepared, and the camera was placed 1.0 [m] from the floor, 3.5[m] from the center of the force plate, and perpendicular to the walking path. In order to collect training and validation data efficiently, videos were shot at 1080p HD/60fps and then converted to images at different frame rates. Participants were asked to walk as usual, and 50 trials were filmed during the sixth step, the stance phase of one gait, when the volunteers were walking normally and the left foot on the front side touched the force plate.

C. Deep Learning Models Creation Methods

Preprocessing is applied to the data obtained from experiments to create training data and validation data. The acquired walking videos are converted into images for each frame rate using the Python module OpenCV. Although the obtained image is a color image, it is converted to a monochrome image to reduce the influence of clothing color on the estimation and converted to 40 x 40 pixels by the bilinear interpolation method. Only the stance phase is extracted by checking the image and matching the time when the foot touches the force plate with the time when the ground reaction force value begins to output due to the foot touching the force plate. The ground reaction force values are normalized by the respective body weight to eliminate differences in values due to body weight and are set to true values. After that, 3D data for input to deep learning is created using the image at the time of the estimation and the images from four images before that time. The number of outputs of the deep learning model are 150, ranging from 0.01 to 1.50 in increments of 0.01. Of the data from the five volunteers, four are used as training data and one as validation data, and the training data is created so that all volunteers become validation data. The number of training and validation data is shown in Table 1 because the number of acquired images is different for each experimental collaborator. the structure of the 3D CNN is as described in Section 2-1, and the parameter values are shown in Table 2 after a trial-and-error process. The deep learning model is created using the deep learning library Keras with reference to the Keras Documentation [8]. EarlyStopping was used as censoring condition, the training error was used as the monitor, and auto was used as the mode. The deep learning model is created by training on the created training data and is terminated by the censoring condition, and estimation is performed on the validation data.

III. RESULTS

Fig. 1 shows the correlation between the estimated value for the 2,013 voxel data of the deep learning model III and the true value that was estimated most accurately. The values estimated using the deep learning model are shown on the horizontal axis as estimated values, and the values obtained using the force plate and normalized by body weight are shown on the vertical axis as true values. Some voxel data are estimated with good accuracy when the true value is larger than 0.70, but not when the value is smaller than 0.70.

Table III shows Pearson's correlation coefficients between the estimated and true values for each deep learning model, the average mean absolute error calculated by multiplying the body weight by the ground reaction force value [N], and the ratio of the mean absolute error to the body weight. The average Pearson's correlation coefficient for the deep learning model was 0.59, showing no improvement in accuracy. Even the deep learning model with the smallest mean absolute error had an error of 10% relative to body weight, and the average mean absolute error for all deep learning models was 15% error relative to body weight, with no improvement in accuracy.

IV. DISCUSSION

In this report, the target estimation accuracy was set at 5% of error to body weight, but the accuracy was 15% error to body weight, showing no improvement in accuracy. Fig. 2 shows the accuracy percentage of correct answers for the training data and the correct answers are shown on the vertical axis, and the accuracy percentage of correct answers for the validation data during the training of the deep learning model III. The percentage of number of epochs is shown on the horizontal axis. The graph shows that the rate of correct answers for the training data improves with each successive training, but the rate of correct answers for the validation data does not. This trend was observed for all deep learning models. This is thought to be caused by overlearning. There are two possible causes of overlearning. The first is the lack of training data, which is a sufficient cause since the current training data are only about 8,000 images each, and we expect improvement by increasing the training data through future experiments. However, to obtain training data from experimental data, a huge amount of experiments must be conducted, and it is difficult to increase training data from experiments because the burden on volunteers, time, and cost are too great. Furthermore, there is no publicly available data set that can be used. Second, there is a bias in the training data. Table 4 shows the percentage of training data per estimation interval for each deep learning model. For all deep learning models, the proportion of training data in the interval between 0.70 and 1.20 accounts for about 80% of the training data, indicating that the training data are biased. In Fig. 1, it can be read that the model is able to estimate in the range where the estimated value is greater than 0.70, but not in the range where the estimated value is smaller than 0.70. This suggests that the accuracy did not improve due to bias in the training data. On the basis of these results, we expect that the number and bias of the training data are problematic. Therefore, if the same amount of training data can be generated for all intervals, such as by expanding the data only

TABLE I NUMBER OF TRAINING AND VALIDATION DATA.

Deep learning models number	I	II	III	IV	V
Training data	B,C, D,E	A,C, D,E	A,B, D,E	A,B, C,E	A,B, C,D
Number of voxel data for the training data.	8252	8046	8255	8153	8366
Validation data	A	B	C	D	E
Number of voxel data for the validation data	2016	2222	2013	2115	1902

TABLE II 3D CNN LEARNING CONDITIONS.

		Set value
Convolution layer	Filter size	5×5×2
	Stride	1
	Channels	256
Pooling layer	Filter size	5×5×2
	Stride	1
Dropout		0.3
Fully connected layer		128
Batch size		100
Epoch		500

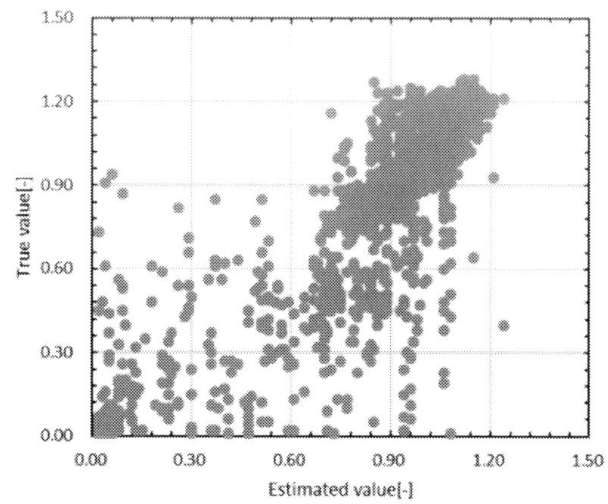

Fig.1 Normalized ground reaction force estimates versus true values.

TABLE III RESULTS FOR ALL DEEP LEARNING MODELS.

Deep learning models number	I	II	III	IV	V	Average
Pearson's correlation coefficient	0.65	0.78	0.85	0.02	0.63	0.59
Mean absolute error [N]	105.5	101.5	63.5	120.3	69.2	92.0
Mean absolute error for body weight [%]	16	14	10	19	15	15

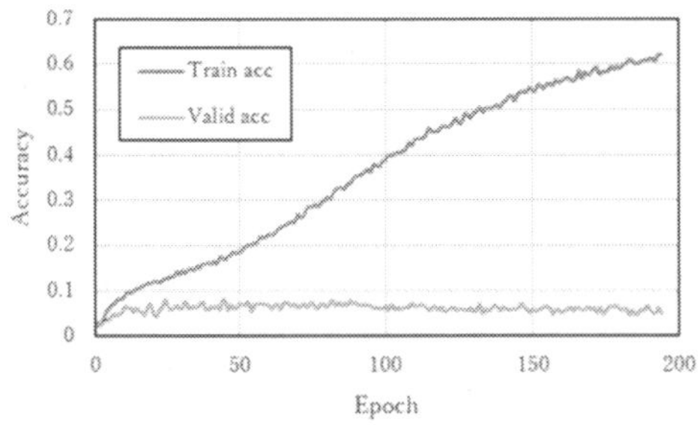

Fig.2 Accuracy rates of training and validation data for deep learning model III.

generated for all intervals, such as by expanding the data only for the intervals where the amount of data was small, the accuracy can be expected to improve.

V. CONCLUSION AND FUTURE WORK

In this report, we examined how to improve the accuracy of the ground reaction force estimation algorithm using only the RGB camera for sensing, which is the proposed method. The method of creating training data was changed, that is voxel data including images at the time of estimation and images up to four images before that time were created and used as training data. No improvement in accuracy was found. It is suggested that overlearning occurs during the training of any deep learning models. We suppose that the overlearning is due to the small amount of training data and bias. Therefore, creating a large amount of unbiased training data is expected to eliminate overlearning and improve accuracy. Another improvement is to incorporate a layer of recurrent neural network into the deep learning model used, in addition to CNN, so that time factor can be learned and accuracy can be improved.

In the future, our aim is to develop a system that can capture images and estimate three directions ground reaction forces using only a tablet device. If this is realized, it will be possible to evaluate gait on a daily by observing ground reaction force waveforms, which will support people to be aware of gait improvement and contribute to extending healthy life expectancy.

REFERENCES

[1] J.Perry, and J. M. Burnfield, GAIT ANALYSIS Normal and Pathological Function, Ishiyaku Publishers, Inc., pp. 243-249, 2007, (in Japanese)

[2] A. Isshiki, Y. Inoue, K. Shibata, and M. Sonobe, "Estimation of Floor Reaction Force During Walking Using Physical Inertial Force by Wireless Motion Sensor," HCI Int'l, vol. 714, pp. 249-254, May 2017, DOI: 10.1007/978-3-319-58753- 0_37, 2017, pp.22-33, ISSN:1348-711

[3] D. Liu, M. He, M. Hou, and Y. Ma, "Deep learning based ground reaction force estimation for stair walking using kinematic data, Measurement," volume 198, July 2022, 111344

[4] S. Sakamoto, D. Owaki, and M. Hayashibe, "Ground Reaction Force Estimation from EMG and IMU Sensor Using Recurrent Neural Network," The Japan Society of Mechanical Engineers Tohoku Branch 55th Annual Meeting and Lecture, Section ID: 108_paper, 2020, (in Japanese)

[5] K. Yagi, Y. Sugiura, K. Hasegawa, and H. Saito, "Gait Measurement at Home Using a Single RGB Camera," Gait&Posture Volume 76, February 2020, Pages 136-140

[6] OpenPose, https://cmu-perceptual-computing-lab.github.io/openpose/web/html/doc/, 2023.10.13

[7] T. Mochizuki, and K. Shibata, "Estimation of Floor Reaction Forces by Convolutional Neural Network Using Walking Image without Depth Information : Evaluation of Generalization Ability," 2023 JSME Information, Intelligence and Precision Equipment Division, IIPB-4-12, 2023, (in Japanese)

[8] keras Documentation, https://keras.io, 2023.10.13

TABLE IV PERCENTAGE OF TRAINING DATA PER ESTIMATION INTERVAL.

Estimation interval	0.01~0.10	0.11~0.20	0.21~0.30	0.31~0.40	0.41~0.50	0.51~0.60	0.61~0.70	0.71~0.80	0.81~0.90	0.91~1.00	1.01~1.10	1.11~1.20	1.21~1.30	1.31~1.40	1.41~1.50
Deep learning model I	5.7%	2.2%	2.1%	2.3%	2.8%	2.9%	3.3%	11.2%	26.4%	17.1%	11.1%	11.8%	1.0%	0.0%	0.0%
Deep learning model II	6.1%	2.0%	1.9%	2.1%	3.2%	3.0%	3.2%	11.6%	21.9%	16.0%	12.4%	15.3%	1.4%	0.0%	0.0%
Deep learning model III	5.7%	2.1%	2.0%	2.2%	3.0%	2.7%	3.4%	11.1%	25.9%	17.4%	11.8%	12.2%	0.6%	0.0%	0.0%
Deep learning model IV	6.3%	2.0%	1.7%	2.0%	3.0%	2.9%	3.2%	12.6%	23.4%	14.6%	13.0%	14.3%	1.1%	0.0%	0.0%
Deep learning model V	6.0%	2.1%	2.2%	2.3%	3.4%	2.6%	2.7%	9.0%	26.7%	16.8%	11.0%	13.9%	1.3%	0.0%	0.0%

Dataset, Usability and Process - Developing an Interdisciplinary, Multi-modal Data Collection Tool and Platform for a Rare Disease

Sinéad Impey, Jonathan Turner, Frances Gibbons, Anthony Bolger, Gaye Stephens, Lucy Hederman, Ciara O'Meara, Ferran De La Varga, John Kommala, Matthew Nicholson, Daniel Farrell, Emmet Morrin, Miriam Galvin
ADAPT Centre
Trinity College Dublin
Dublin, Ireland
e-mail: sinead.impey@adaptcentre.ie, jonathan.turner@adaptcentre.ie, frances.gibbons@adaptcentre.ie, anthony.bolger@adaptcentre.ie, gaye.stephens@tcd.ie, hederman@tcd.ie, ciara.omeara@adaptcentre.ie, ferran.delavarga@adaptcentre.ie, john.kommala@adaptcentre.ie, matthew.nicholson@adaptcentre.ie, daniel.farrell@adaptcentre.ie, emmet.morrin@adaptcentre.ie, galvinmi@tcd.ie

Mark Heverin, Éanna Mac Domhnaill, Robert McFarlane, Dara Meldrum, Deirdre Murray, Orla Hardiman
Academic Unit of Neurology
Trinity College Dublin
Dublin, Ireland
e-mail: mark.heverin@tcd.ie, amacdomh@tcd.ie, macfarlro@tcd.ie, meldrumd@tcd.ie, dmurray1@tcd.ie, hardimao@tcd.ie

***Abstract*—Large data sets are required to understand disease progression, investigate treatment options and discover potential cures in rare neurological conditions such as Amyotrophic Lateral Sclerosis (ALS). Generating large data sets for such rare neurological conditions requires the participation of multiple clinical sites. The Precision ALS project is a partnership between multiple clinical sites and industry partners across Europe that seeks to collect and analyse multi-modal data collected from participants with the disease. In this paper, we describe the development of a data collection tool that allows for the collection and integration of data collected at these multiple sites. We focus particularly on the requirements gathering method, which was divided into three pillars: Dataset, Usability and Process. The data collection tool runs on an Android tablet and is now in use enabling collection of data from across Europe for the Precision ALS project.**

Keywords-amyotrophic lateral sclerosis; motor neurone disease; agile development process; requirements gathering; data integration.

I. Introduction

Amyotrophic Lateral Sclerosis (ALS) is an incurable progressive neurodegenerative disease responsible for up to 10,000 deaths per year in Europe; it is the most common form of the motor neuron diseases [1][2]. Most (> 90%) cases of ALS have no known cause [3][4]; the remaining cases have a genetic cause [5][6]. Collaboration between clinicians and data scientists is required in the collection, curation and analysis, including by machine learning methods, of large multi-modal data set to understand the disease and its heterogeneity [7]. However, generating such large datasets for a rare disease like ALS can be challenging due to the low numbers of affected individuals. In response to this challenge, the Precision ALS (P-ALS) project [7][8] was initiated as a partnership between nine clinical sites, interested industry partners and technical researchers. The project aims to develop a data collection tool and an interdisciplinary data platform to gather, share and analyse multi-modal data. Following this introduction, the rest of this paper is organized as follows. Section II describes the methods used. Section III describes the early results from this work. Section IV discusses our conclusions and plans for future development. The acknowledgement and references close the article.

II. Method

Development of the data collection tool was in two streams: requirements gathering and refinement, and application development and refinement. Requirements gathering focused on the needs of clinicians, data collectors and analysts; application development was the domain of the technical team. Refinement of requirements and of the developed application was dependent on effective two-way communication between the clinical and technical groups. To ensure effective communication during requirements gathering, visits from members of the clinical and development team to data collection sites took place, to meet with clinicians, researchers and data collectors at each site.

A. *Requirements gathering*

To ensure that the needs of the project stakeholders were met, a requirements process was developed which divided the requirements gathering for the data collection tool into three pillars as shown in Fig. 1. Pillar 1, "Dataset", collected content requirements for the data collection tool, i.e., the fields to be collected. Pillar 2, "Usability", focused on requirements for the workflow through the data collection tool to maximise data collection efficiency and accuracy. Pillar 3, "process", investigated requirements relating to the flow of data into the collection tool from the various data sources that contribute to the final dataset.

1) *Pillar 1: dataset*

In Pillar 1 of the requirements gathering, fields to be used in the data collection were identified. Expert sources used for identification of these fields included clinicians specialising in ALS, data analysts, and individual partner sites who had been collecting data on ALS patients for some years. Following identification of fields needed, options within those fields were identified for inclusion on a paper worksheet. This worksheet was used as a prototype for the tool. For example, participants were to be questioned about their history of taking ALS symptomatic medications; a list of possible medications was included in the data collection tool as options from which the data collector could select appropriately.

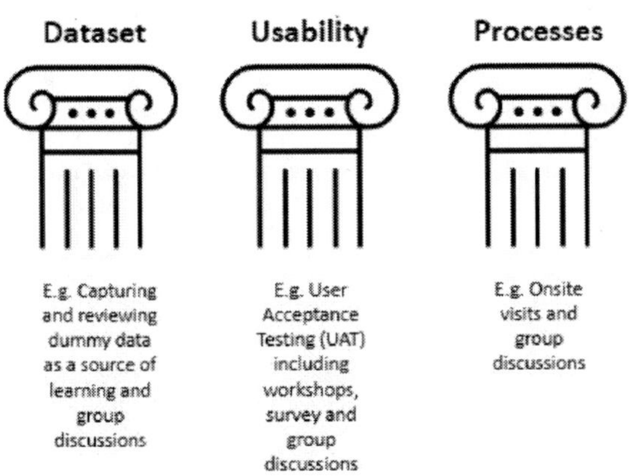

Figure 1. The three pillars of requirements gathering.

2) *Pillar 2: usability*

The data collection tool is intended to be used in a variety of settings, including face-to-face with participants and/or their carer(s), in telephone interviews, or for data entry from existing records. In each scenario, the tool must be usable efficiently and accurately by the data collector. With no existing data collection process in place, usability of the tool was focused on clear presentation of data items and ease of data entry.

Representatives from each site, who were planning to use the tool for data collection, were given a complete walkthrough of the collection tool, at a single meeting involving all data collection sites. Users then had the opportunity to explore the tool independently, with members of the development team available to answer any questions and to capture verbal feedback. Following this familiarization process, a user survey was completed by each site member, 13 users in total, using version 3 of the Post-Study System Usability Questionnaire (PSSUQ) [9]. It is intended that the tool will be further refined over the period of its development and use, with changes in usability measurabl-e by repeat use of the questionnaire. Results of the baseline PSSUQ are shown in Table 1. Possible scores range from 1 (best) to 7 (worst).

TABLE I. RESULTS OF THE BASELINE PSSUQ.

Question number	Question text	Average Score
1	Overall, I am satisfied with how easy it is to use the system	2.2
2	It was simple to use this system	2.2
3	I was able to complete the tasks and scenarios quickly using this system	2.4
4	I felt comfortable using this system	2.6
5	It was easy to learn to use this system	2.3
6	I believe I could become productive quickly using this system	2.0
7	The system gave error messages that clearly told me how to fix problems	3.1
8	Whenever I made a mistake using the system, I could recover quickly and easily	2.7
9	The information (such as SOP) provided with this system was clear	2.3
10	It was easy to find the information I needed	2.0
11	The information was effective in helping me complete the tasks and scenarios	2.3
12	The organization of information on the system was clear	1.9
13	The interface was pleasant	2.0
14	I like using the interface of this system	2.2
15	The system has all the functions and capabilities I expect it to have	2.7
16	Overall, I am satisfied with this system	2.0
	Overall PSSUQ score	2.3
	System usefulness	2.3
	Information quality	2.4
	Interface quality	2.3

3) Pillar 3: process

The process pillar is concerned with understanding the current processes and associated actors, both human and technological involved in data collection. The purpose of this is to gain an understanding of how the proposed technology may impact the current processes or where changes are likely. This pillar ensures that the data collection process required by the tool does not impose inefficient or impractical working practices on the data collection sites. Existing data collection processes at data collection sites were examined for roles of actors involved in the collection process, in particular: the role of data collectors; registration of participants; clinical coding systems used; information systems used; and remote monitoring of participants. The collected data is personal and sensitive, and each collection site was required to follow their local processes for ethics approval, data protection impact assessment and to sign a data transfer agreement. It was for each site to ensure that they comply with their local data protection laws. Oversight of these processes was coordinated at the top level of the project via a team that included data protection experts.

The three pillars, although discussed separately, in practice were reviewed together. To do this, two project researchers visited each site. This allowed the researchers to discuss and compare findings. Each site was encouraged to include as many stakeholders as they wished in these visits, but it was imperative that the data collectors were available. Initial discussion focused on the data collection process proposed by the site which included location of such data as medical records, current studies or discipline-specific databases.

To understand the variables contained in the worksheet and the interface design, the data collector was asked to take part in a mock interview. For this interview, one researcher acted as the participant and the other researcher took notes. The data collector at the study site used the collection tool to capture participant responses and was encouraged to discuss their thoughts on the interpretation of the question, what type of answer they might expect and how the response could best be captured in the tool.

Information gathered during the site visits was discussed between the researchers post interview and recorded. These were brought to the wider project group for a decision or further discussion. It is expected that some of these findings could be incorporated into standard operating procedures.

B. Application development

Development of the data collection tool was carried out in-house in the ADAPT Centre [10], with the development team based at Trinity College Dublin. This team met regularly with clinicians and data analysts to ensure that the developed tool met their needs.

A tablet-based application approach was chosen to ensure portability and to enable operation without a working internet connection. It was decided that dedicated devices would be provided to the sites for data collection, which would be managed remotely This application is deployed via a mobile device management solution to minimise security and device management concerns. Using Android with Mobile Device Management software provides the mechanism to distribute private apps and client certificates, and allows for restricted and secure access to the server. Android provides a more open and accessible development platform than Apple and iOS, which does not provide a distribution mechanism for the small scale required. The application was developed by the in-house team using the Kotlin programming language [11] in Android Studio [12]. The data collection form structure is configured using a metadata driven approach, allowing easy updates without the need to modify the application code itself. Development followed a lightweight Agile [13] approach with regular prototype releases to project stakeholders.

III. Results

Use of the three pillars for requirements gathering was successful. From Pillar 1, dataset, an agreed data set was identified. Pillar 2, usability, identified functions to improve engagement; results of the user survey are shown in Table 1. Pillar 3, process, identified the people and systems currently used at data collection sites and how these actors could be replicated in the collection tool. A sample data collection page is shown in Fig. 2. This also lists, on the left, the full set of pages available in the collection tool. The data collection tool contains 15 pages, each focusing on a particular division of data to be collected, for example 'Smoking and Alcohol' use, or 'Socio-Economic Details'. This allows for some pages to be skipped when not appropriate, e.g., during a repeat data collection encounter when the focus of the data collector is on fields that may have changed since the last encounter, such as clinical progression or resource use.

The developed version of the data collection tool was first used in a clinical setting in August 2023, at one site. Further sites started data collection in September 2023.

IV. Conclusions and Future Work

A set of requirements for the development of a data collection tool can be constructed from the requirements of different groups of interested parties (clinicians, analysts, industry partners), with the success of the developed tool dependent on regular communication between these parties and the technical development team. The tool can incorporate requirements from existing data collection practices at individual partner sites and new requirements elicited as part of the requirements gathering process.

Future work has three main strands. Development and refinement of the data collection tool will continue, with the knowledge and feedback gained from its use in the field informing this. Following the successful collection of data from participants in the P-ALS research project, development of the data platform infrastructure will commence using a similar development process, i.e., requirements gathering from clinicians, researchers, data

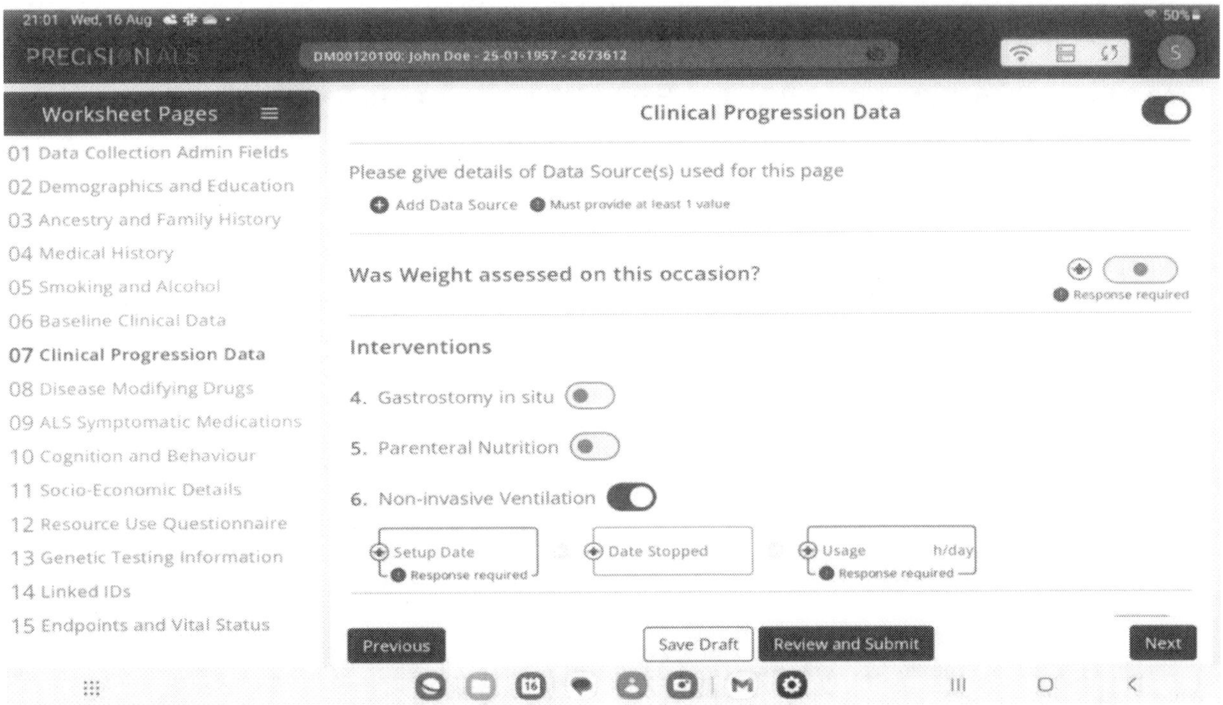

Figure 2. An example page from the data collection tool.

development team of the data platform structure required.

Data platform infrastructure development will include ensuring that data from multiple modalities, e.g., medical imaging, wearable devices, can be imported into the platform and made available for analysis. Development of the platform will be described in future -publications. Finally, once data of sufficient quantity and quality is available, analyses can be performed on the data. Work is underway to ensure that academic analysts and industry partners have the opportunity at an early stage to describe questions that they may wish to answer, to ensure that the data platform enables the ability to answer these questions. These questions include the costs of the disease, both to society and to families; if incidence of ALS is higher for particular occupations; and whether time to disease progression events can be predicted from early information on an individual.

ACKNOWLEDGMENT

This work is funded by the Science Foundation Ireland grant SP20/SP/8953. OH's work is also funded from grants 16/RC/3948 and 13/RC/2106_P2.

REFERENCES

[1] M. Ryan, M. Heverin, M. A. Doherty, N. Davis, E. M. Corr, et al. "Determining the incidence of familiality in ALS: a study of temporal trends in Ireland from 1994 to 2016," Neurology Genetics. 2018;4:e239.

[2] C. A. Johnston, B. R. Stanton, M. R. Turner, R. Gray, A. H. Blunt, et al. "Amyotrophic lateral sclerosis in an urban setting: a population-based study of inner city London". Journal of Neurology. 2006;253:1642–3.

[3] National Institute of Neurological Disorders and Stroke. *Amyotrophic Lateral Sclerosis (ALS) Fact Sheet*. [Online]. Available from: www.ninds.nih.gov [Accessed 2023.08.21]

[4] ALS Association. *Understanding ALS*. [Online]. Available from: https://www.als.org/understanding-als [Accessed 2023.08.21]

[5] S.A Goutman, O. Hardiman, A. Al-Chalabi, A. Chió, M. G. Savelieff, et al. "Recent advances in the diagnosis and prognosis of amyotrophic lateral sclerosis". The Lancet. Neurology. 21 (5): 480–493, May 2022, doi:10.1016/S1474-4422(21)00465-8. PMC 9513753. PMID 35334233

[6] MedlinePlus. *Amyotrophic lateral sclerosis*. [Online]. Available from: https://medlineplus.gov/genetics/condition/amyotrophic-lateral-sclerosis [Accessed 2023.08.21]

[7] R. McFarlane, M. Galvin, M. Heverin, É. Mac Domhnaill, D. Murray, et al, "PRECISION ALS—an integrated pan European patient data platform for ALS," Amyotrophic Lateral Sclerosis and Frontotemporal Degeneration. 2023. 24:5-6, 389-393, DOI: 10.1080/21678421.2023.2215838

[8] Precision ALS. *Precision ALS.* [Online]. Available from: www.precisionals.ie [Accessed 2023.08.21]

[9] J. R. Lewis. "IBM computer usability satisfaction questionnaires: Psychometric evaluation and instructions for use". International Journal of Human–Computer Interaction. 1995. 7:1, 57-78, DOI: 10.1080/10447319509526110

[10] Adapt Research Centre. *ADAPT: The Global Centre of Excellence for Digital Content and Media Innovation.* [Online]. Available from https://www.adaptcentre.ie/ [Accessed 2023.08.31]

[11] JetBrains. *Kotlin.* [Online]. Available from: https://kotlinlang.org/ [Accessed 2023.08.21]

[12] Google LLC. *Android Studio.* [Online]. Available from: https://developer.android.com/studio [Accessed 2023.08.21]

[13] Agile Alliance. *What is Agile?* [Online]. Available from: https://www.agilealliance.org/agile101 [Accessed 2023.08.21]

Medication Adherence Prediction for Homecare Patients, Using Medication Delivery Data

Ben Malin
Dept. Electronic and Electrical Engineering
Brunel University London
London, United Kingdom
e-mail: ben.malin@brunel.ac.uk

Tatiana Kalganova
Dept. Electronic and Electrical Engineering
Brunel University London
London, United Kingdom
e-mail: tatiana.kalganova@brunel.ac.uk

Ejike Nwokoro
Patient Insights and Data Strategy Unit
HealthNet Homecare
London, United Kingdom
email: ejike.nwokoro@healthnethomecare.co.uk

Joshua Hinton
Patient Insights and Data Strategy Unit
HealthNet Homecare
London, United Kingdom
email: joshua.hinton@healthnethomecare.co.uk

***Abstract*—This study aims to predict the risk of medication nonadherence for patients who are newly enrolled into a medication delivery homecare service – an insight that can underpin the design of more impactful patient support programs for patients with long term conditions. In the context of this study, we have defined a nonadherent patient as someone without any prescribed medication available across the month. This is calculated using medication delivery confirmation and prescription data. Convolutional Neural Networks (CNN) and Random Forest (RF) networks are used for this study, with the former shown to be our best-performing model, achieving an 82.8% Area Under the Curve (AUC) on a subset of the patient population who have been on service for 3 to 4 months. When testing the model on the entire patient population (regardless of how long they have been on service), and by using cross-validation, the AUC improves to 97.4%. The methodology that is applied in our study is novel based on three distinct factors: (1) prediction that is based on a novel visualization of 12 months of patient medication delivery data, (2) taking into consideration the temporal patient communications as well as the possibility of patient stockpiling of prescribed medication and (3) the service level i.e., level of nurse support received by the patient. We find that the inclusion of temporal patient communication data into our analysis improves both the AUC and the nonadherence prediction precision in the CNN model (0.7% and 19.4% respectively); a similar improvement in AUC and prediction precision is not seen in the RF model. The CNN model is therefore identified as the appropriate model for our use case. Furthermore, our results support the claim that temporal communication data are relevant datapoints for predicting adherence in a network that is better-suited to time-series data.**

Keywords- medication adherence; CNN; RF, healthcare; homecare; adherence prediction

I. Introduction

Medication adherence is a vital aspect of a patient's treatment journey, with adherence being linked to positive disease outcomes as well as lowering the burden on the healthcare provider. This is evidenced through global estimates of nonadherence causing 125,000 patient deaths per year, as well as $100 billion in preventable medical costs [1]. Machine Learning (ML) has been utilized effectively in predicting nonadherence, with the intent of providing these potentially nonadherent patients with the support necessary to keep them adherent – reducing risk to health, as well as future treatment cost [2]. It is however noted that different therapy areas and medications have different levels of burden, complexity, as well as different rates of nonadherence [3][4], and as such may require tailored interventions. In our study, the proposed use case for predicting adherent and nonadherent behavior is across a diverse range of therapy areas including, dermatology, gastroenterology, rheumatology and respiratory. The goal is to improve patient wellbeing across all therapy areas through predictive adherence, whilst acknowledging the role that a variety of clinical and non-clinical factors can play in influencing the chances that a patient will adhere to their prescribed therapy. Our study focuses on patients who are newly enrolled to the homecare delivery service.

This paper is structured into the following sections to provide insight into the work conducted. In "Background and related work", existing literature is reviewed for ML networks that have been previously used for adherence prediction, along with the data inputs and adherence metrics utilized. "Design decisions" explains why the identified approaches are most suitable for this study, along with any other modifications for our use case, "Data source and processing" delves into how these design decisions have been implemented with the data available to us. The "Methodology" section will explain our implementation of different ML models. In "Results", we will present our findings, followed by "Results comparison" to help contextualize our results against other studies. The "Discussion" section will provide our analysis of the results.

Finally, the "Conclusion" will summarize these insights, as well as provide suggestions for future research.

II. BACKGROUND AND RELATED WORK

The approaches for assessing and improving medication adherence differ greatly, with the main distinctions being the type of ML models that are utilized and the range of data types and/or adherence metrics incorporated into such models.

Adherence can be difficult to define, as different diagnoses have differing levels of repercussions for varying degrees of nonadherence [3][4]. However, the most used metrics for adherence tend to be the use of prescription refill data or a patient-reported adherence score, based on a questionnaire [5]–[11] . Both metrics typically use a binary label for whether a patient is adherent or not, with the prescription refill-based approach often defining a patient as adherent if they have a Proportion of Days Covered by medication (PDC) above 80% of the total predetermined timeframe [5][7][12]. However, it is important to note that this is an indirect measure of adherence, as it cannot be known whether the patient consumes their prescribed medication, only whether they have received it. The alternative patient-reported adherence score definition is typically based on a series of scored questions, and it is common for nonadherence to be defined as below 80%-85% of the maximum achievable score [6][9][10].

It is therefore the chosen adherence metric that will determine what the model will predict, i.e., whether it predicts if the patient will have medication for 80% of the next month, or whether the patient will respond positively to adherence-related questions resulting in an adherence score above 80-85%.

The initial phase of our study entailed a review of previous research within the predictive adherence space. As part of this review process, data types used by the various predictive methods are broken down into groupings. These groupings are shown in Table 1.

Table 2 compares the data inputs across various research studies along with their use case and the prediction method that was used. The most common ML architectures used for adherence prediction tend to be decision-tree (DT) based, in particular, Random Forest (RF) and other decision-tree methodologies. Other methodologies have also been previously considered such as Long Short-Term Memory (LSTM), k-Nearest Neighbor (KNN), Logistic Regression (LR), Gradient Boosting (GB) and Artificial Neural Networks (ANN). None of the reviewed studies (Table 2) use Convolutional Neural Networks (CNNs) despite how commonplace they are in other areas [13][14]. Similarly, Long Short-Term Memory (LSTM) models appear to be unpopular in adherence prediction, despite having achieved better performance than RF models in some cases [2].

One benefit of the CNN and LSTM models is that they can easily accommodate windowed time-series data to the network in a way that RF cannot, as RF is reliant upon

TABLE I. PREDICTIVE ADHERENCE DATA INPUT GROUPINGS

Patient Profile Data	**Medication Supply Data**	**Communications Data**
Demographics	Adherence levels	Communications with medication provider
Comorbidities	Medication complexity	
Service status	Medication supply	
	Prescription details	

independent feature input [14]–[16]. This results in separate features in an RF network detailing characteristics, such as adherence on specific days, as opposed to a single sequential data input [5][17]. Thus, RF networks can be used successfully, but may be inferior to LSTMs and CNNs for time-series prediction tasks. CNNs have been used successfully for time-series forecasting and prediction, outperforming LSTMs, and other models [18]–[21]. Utilizing historical medication data for the prediction of adherence is a comparable use case, and another reason for the evaluation of CNNs in this field.

Table 2 shows that patient profiling and medication delivery data are often used for adherence prediction, with some studies deeming the latter as having a stronger influence on such predictions [2].

To our knowledge, little research has been done to examine the extent to which time-series data, outside of medication stock, is relevant to adherence behavior. This is where the use of patient communications data as an additional variable in predicting adherence behavior is novel, as it provides information to the model regarding a patient's level of interaction with their prescriber and/or medication delivery service provider.

III. DESIGN DECISIONS

The chosen adherence metric for our study is PDC, following the trend demonstrated by other studies in Table 2. There are several additional reasons for this, including, that the PDC metric has been advocated for by various bodies (e.g., the Pharmacy Quality Alliance (PQA)) as the preferred quality indicator for estimating adherence to therapies for chronic diseases [22].

Additionally, PDC captures the number of days the medication should last, rather than the number of days the medication is in a patient's possession. Hence it makes no difference if the patient collects the medication early. This capability means the metric lends itself well to the medication delivery data used in our study, as we can calculate the number of days that each prescription should last. Importantly however, unlike most of the studies included in our initial literature review, we deem 100% of days covered as adherent, and any value less than this as nonadherent. The

TABLE II. MACHINE LEARNING MODEL EVALUATION FOR ADHERENCE PREDICTION

Author, year	**Therapy Area**	**Adherence Metric**		**Architecture**	**Input Variables**		
					Patient Profile	**Medication Supply Data**	**Communications**
Franklin et al., 2015 [23]	Cardiovascular disease	Binary	>PDC 80 – Medication Dispensation Date	Group trajectory modelling	✓	✓	
Lucas et al., 2017 [24]	Cardiovascular disease	Binary	>PDC 80 – Medication Prescription Date	RF	✓	✓	
Kumamaru et al., 2018 [12]	Cardiovascular disease	Binary	>PDC 80 – Medication Dispensation Date	LR	✓	✓	
Haas et al., 2019 [6]	Fibromyalgia	Binary	Self-reported	RF	✓	✓	
Kim et al., 2019 [9]	Smoking addiction	Binary/ Tertiary	Patient estimated adherence	DT	✓	✓	
Galozy et al., 2020 [5]	Hypertension	Binary	>80% PDC – Medication Refill Date	RF, LR, GB, KNN	✓	✓	
Gao et al., 2020 [25]	Hypertension	Binary	>PDC 80 – Medication Prescription Date	DT	✓	✓	
Koesmahargyo et al., 2020 [17]	Diverse – predominantly mental diagnoses	Binary	>80% Recommended Daily Medication Consumption	GB	✓	✓	
Wang et al., 2020 [10]	Crohn's disease	Binary	Self-reported	SVM, LR	✓		
Wu et al., 2020 [11]	Type 2 Diabetes	Binary	>PDC 80 – Medication Prescription Date	SVM, KNN, DT, Ensemble	✓	✓	
Gu et al., 2021 [2]	Diverse diagnoses	Binary	Medicine taken on time	LSTM, RF, GB		✓	
Kharrazi et al., 2021 [8]	Diverse diagnoses	NA – predicting hospitalizations		LR	✓	✓	
Li et al., 2021 [26]	Hypertension	Binary	Medicine taken on time	LR, DT, ANN, RF	✓		
Hsu et al., 2022 [7]	Cardiovascular disease	Binary	>80% PDC – Dispensation Date	LSTM		✓	
(This Work, 2023)	Diverse diagnoses – asthma, dermatitis, psoriasis and more	Binary	100% PDC – Medication Delivery Date, >80% PDC – Medication Delivery Date	CNN	✓	✓	✓

motivation behind adopting this strict approach is the diversity of therapies prescribed to the patients included in our study and the variability in the ways that nonadherence can affect different patients and diagnoses [4][27].

The medication delivery frequency and stock level for the patients included in our study is driven by the patient's prescription and provides the recommended quantity of days of medication. It is important however, to note that patients who have their medications delivered direct to home can request more than their usual level of stock (for example just before a holiday), leading to a deviation in their standard delivery frequency. This behavior is commonplace for patients with a chronic disease [28][29]. It is therefore necessary to consider whether a patient has previously 'stockpiled' medication before they are deemed as nonadherent for failing to take delivery of their medication. While there are medication adherence studies based on prescription dispensing dates, to the best of our knowledge, ours is the first study predicting adherence using medication delivery confirmation data and taking into consideration the potential of stockpiling [23][19].

IV. Data Source and Preprocessing

Patients included in our study are those diagnosed with long-term conditions and who have been receiving direct to home delivery of their medication as well as nurse support for medication self-administration at home from a clinical homecare provider (HealthNet Homecare Ltd). The dataset contains, but is not limited to, demography, length of time on homecare service (LOS), primary diagnosis, medication delivery confirmation, nurse visit confirmation and whether the patient receives enhanced nurse support. Such enhanced nurse support is used to aid medication adherence.

Patients are excluded if they have finished their treatment, or if they have not had a medication delivery within the last 13 months. The medication delivery confirmation data contains the date-time at which each patient receives their medication, as well as the number of days that the delivered medication covers them for. Thus, any periods where the recommended medication delivery frequency is insufficient can be calculated. Additionally, periods where the patient received additional (i.e., extra) deliveries are accounted for, allowing for medication stockpiling, which is common in chronic disease management [29].

The medication delivery confirmation across each patient's latest month is calculated, and any period with a lapse of medication delivery confirmation frequency in this time period designates that patient as nonadherent. This label is used as the target variable for the model. The medication delivery confirmation data over the 12 months prior to the target variable month is used for model training and inference, as with this it is possible to determine whether the model accurately predicts nonadherence in the target variable month.

Figure 1 shows how the 12 months of medication delivery confirmation is visualized, with the horizontal axis representing time. The left-most point on the axis representing 13 months before the most recent dataset entry, and the right-most point representing one month before the most recent dataset entry – providing 12 months of patient medication timeline data. These areas are then colored based on the medication stock quantity that the patient should have on each day within this time period, and this representation can be seen in Figure 1a. Dark green represents sufficient medication stock of >31, light green is sufficient stock for 5-31 days, yellow is for 1-5 days, and red means the patient is not in possession of any medication stock. In instances where the patient has been on the service for less than 13 months, the period prior to beginning their treatment is color coded as white.

Additional information can also be encoded into these images, in the form of delivery communications and the patient's enhanced nursing support service status, merging temporal and non-temporal data into a single sample. This is illustrated in Figures 1b and 1c, where these data visualizations show different encoded information for the same patient.

Figure 1b represents a patient's enhanced service status as a solid line across the 12-month period, the color of which is blue for receiving enhanced services and black for not receiving enhanced services (not shown). Figure 1c encodes all delivery communications as colored dots, with the features of the dot representing the type of communication. The color and outline of the dots varies to reflect the medium used for communication as well as whether the communication was inbound or outbound. Thus, creating a unique color scheme for each communication type across our dataset. Where multiple communications occur on the same day, the subsequent dots are placed below the previous communications in chronological order.

V. Methodology

We implemented both RF and CNN models in this study. The RF network is intended to be used as a benchmark, due to its ubiquity for adherence prediction across other studies. The CNN on the other hand, was chosen given its capabilities in time-series prediction, as well as our desire to test its feasibility in adherence prediction whilst utilizing images produced with heterogeneous data [18][20].

Other studies have shown that CNNs can be effectively used on visually represented data that has been generated from both time-series data and tabular data, outperforming the use of the raw numerical data [18][19].

The approach of representing numerical/time-series data into the visual data domain is a common preprocessing technique that is used in signal processing for the improvement of performance [30][31]. Taking inspiration from this domain and translating it to time-series patient data requires novel preprocessing, as techniques that are common in signal processing are incompatible with our data, such as Gramian Angular Summation/Difference Fields, Markov Transition Fields and spectrograms [30]–[32]. These images are automatically generated in a deterministic manner for each patient, using their medication stock level over time. Implementing a deep CNN architecture, Inception v3, on these data visualizations is straightforward and has the potential to learn features that the RF model cannot.

Each experiment is performed five times to reduce the influence of different random initializations. Additionally, 5-fold cross validation tests have been performed when there is

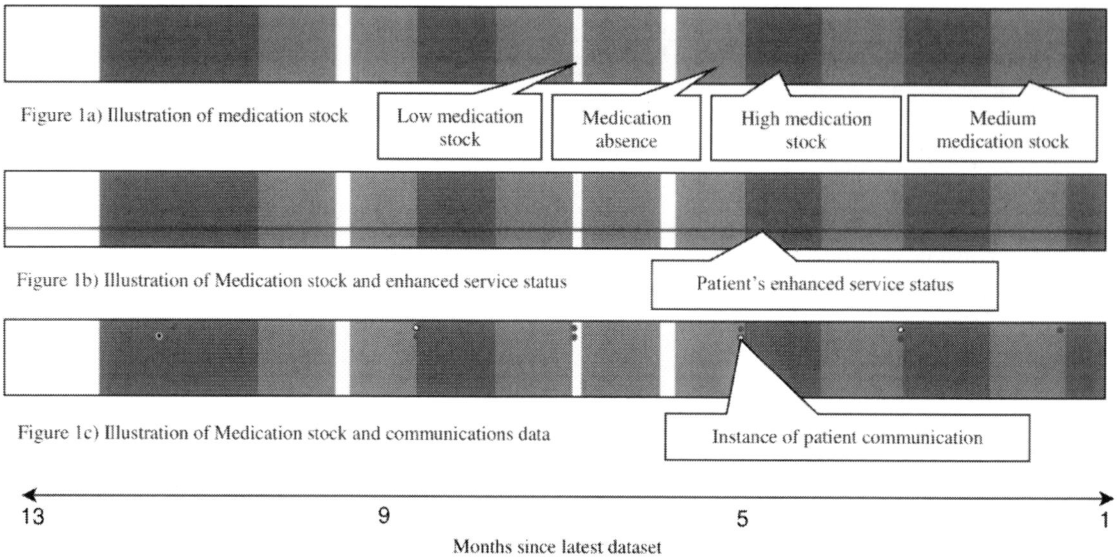

Figure 1 Patient timeline images

no predetermined test set – iterating through which patients are included into the training and testing sets. This validation strategy is commonplace for adherence prediction [7][24]. CNN training was performed using 50 epochs per run with the Adam optimizer.

Additionally, undersampling was trialed for every test, due to the imbalance between the majority and minority class. When undersampling yielded the best result this has been stated. Undersampling is the process of data reduction to distribute the data quantity between two classes more evenly, with the aim of reducing the likelihood of misclassifying the minority class [33]. In our case, this will refer to removal of adherent samples, which comprise approximately 80% of all samples. Multiple studies have successfully used this balancing technique to improve their performance [5][26]. Oversampling is another technique that could be utilized for reducing class imbalance and has been used for adherence prediction previously, though it is out of scope for this study [26].

VI. Model Results

Initial comparative testing was conducted using both RF and CNN architectures. These tests were performed using the entire patient population data (i.e., regardless of how long they have been on the homecare service), using 5-fold cross-validation to determine which patients comprise the test set. For the RF network, each patient's data is inputted as 365 features with each feature containing the number of days' worth of medication stock they should have in their possession. When delivery communications data is utilized, these features are inputted using one-hot encoding to the network, with a feature for each day. The enhanced services feature is a single feature that denotes a binary variable. The data input for the CNN is image-based and Figure 1 is representative of the data variants used for this test. These results can be seen below in Table 3. The RF network attains comparable AUC, but with superior nonadherent precision. Additionally, the CNN attains its best performance through the inclusion of delivery communications data and enhanced service status information, unlike the RF which peaks with just medication delivery confirmation data and enhanced service status information. This could be due to the RF network's inability to process time-dependencies across features, unlike a CNN where our inclusion of delivery communications marginally improves AUC but provides a significant gain to nonadherent precision.

However, the primary objective of this study is to predict the risk of non-adherence for patients who are new to the homecare delivery service. The main reason behind this is that we wish to identify potential nonadherence before it occurs, so that targeted interventions can be initiated. As well as the need to make accurate predictions for patients without requiring them to have been on the service for a long time – during which they may have benefited from greater support.

To achieve this primary objective, a test set was created exclusively of patients within the first 3-4 months Length Of Service (LOS) range. Patients with an LOS under 3 months are excluded due to there being limited data for each patient, after the removal of one month for use as the target variable. None of the same patients/patient data were included in both the training and testing data sets. The training set uses data from patients who had joined the service earlier than (i.e., before) the patients whose data was used for testing. This allows for training samples that are comparable to the testing samples, in terms of LOS, without the inclusion of testing data. Once again, both RF and CNN model architectures were implemented in order to identify the optimal network as well as the optimal data. These results can be seen below in Table 4.

TABLE III. CNN AND RF PRELIMINARY DATA INPUT EVALUATION

Medication Availability	Enhanced Services	Delivery Communications	Mean AUC		Mean Nonadherent Precision	
			RF	CNN	RF	CNN
✓			95.36%	95.70%	88.64%	80.02%
✓	✓		95.40%	95.14%	88.49%	81.31%
✓		✓	94.90%	96.26%	88.13%	87.69%
✓	✓	✓	94.92%	97.40%	88.53%	90.10%

Testing on the predetermined set of patients with 3-4 months LOS results in a substantial decrease in performance, compared to testing against the full population. However, for this use-case, the CNN, compared to the RF model, provides better AUC and nonadherent prediction precision. As is the case when the model utilized the entire patient population data, the CNN with the highest AUC incorporates both enhanced service status data and delivery communications data, whilst the best RF model does not use delivery communications data.

VII. RESULTS COMPARISON

For greater comparability with other studies, a PDC of 80% was used in addition to a PDC of 100%, as this is the most common medication availability adherence measure [8][9][14]. Additionally, 5-fold cross-validation was used to evaluate the dataset – with each patient's sample providing 12 months of data. These results can be seen below in Table 5, where our best-performing CNN and RF models are shown. However, the other studies shown in this table do not define medication availability through delivery data, instead this data is provided through in-person prescription refills. Additionally, the other studies have not factored in medication stockpiling which will impact the adherence dynamics. These distinctions, along with the fact that alternative cohorts of patients likely have different demographics and behaviors, do separate the studies from one another. However, comparisons between the studies can be drawn with these caveats, to compare the use of differing data inputs.

The model with the best AUC in our tests was an undersampled CNN which made predictions by utilizing medication delivery confirmation data, enhanced services status information, and delivery communications data, all formatted into an image, as shown in Figure 1. This model attained the highest AUC when predicting both, whether a patient would have a PDC>80% or PDC of 100% in their latest month. The CNN benefited from the use of random undersampling, whereas the RF network performed worse when undersampled. Additionally, the RF network shown in Table 5 does not utilize delivery communications as this data input led to an AUC reduction.

VIII. DISCUSSION

We investigated various methodologies for predicting nonadherence of patients, across both CNN and RF networks. The methods trialed incorporated differing levels of medication delivery confirmation data, timestamped patient communications and a binary variable representing the level of service that a patient receives. This testing was done on a predefined test set of patients with LOS ranging from 3-4 months, in line with the overarching project objective. It was found that the best-performing model utilized all the available encoded data, giving an AUC of 82.8%, with a PDC requirement of 100%. For the CNN, the greatest performance improvements were attained through the inclusion of both the enhanced services status data, as well as the delivery communications data – in addition to medication delivery confirmation data. These features, when encoded into the images, improved AUC by 0.7% and nonadherent prediction precision by 19.4% from the model without these features, thus supporting the claim that these are relevant datapoints for predicting nonadherence. The RF network attained its best performance without the inclusion of delivery communications data, likely due to RF networks being unable to process time-dependencies across features.

TABLE IV. CNN AND RF 3-4 MONTH LOS TEST SET EVALUATION

Medication Supply	Enhanced Services	Delivery Communications	Mean AUC		Mean Nonadherent Precision	
			RF	CNN	RF	CNN
✓			55.44%	82.12%	38.74%	19.13%
✓	✓		53.64%	68.49%	37.81%	80.33%
✓		✓	54.47%	80.05%	37.01%	40.07%
✓	✓	✓	54.81%	82.84%	37.99%	38.54%

TABLE V. PREDICTIVE ADHERENCE COMPARISON USING PDC

Author, year	Adherence Metric	Input Variables			Validation strategy	AUC	Training Samples	Test Samples
		Patient Profile	Medication Availability Data	Communications				
Lucas et al., 2017 [24]	PDC >80%	✓	✓		30-fold cross-validation	73.60% - 81.00%	134,107	4,624
Kumamaru et al., 2018 [12]	PDC >80%	✓	✓		Logistic Regression	Up to 69.60%	49,745	49,745
Galozy et al., 2020 [5]	PDC >80%	✓	✓		Stratified random split *(undersampled)*	80.30% - 80.70%	15,794	2,787
Gao et al., 2020 [25]	PDC >80%	✓	✓		10-fold random seed	81.00%	5,730	1,908
Wu et al., 2020 [11]	PDC >80%	✓	✓		10-fold random seed	57.70% - 86.60%	401	40
Hsu et al., 2022 [7]	PDC >80%		✓		5-fold cross-validation *(predetermined test set)*	80.50%	90,000	10,096
(This work, 2023)	PDC >80% CNN	✓	✓	✓	5-fold cross-validation *(undersampled)*	97.89%	5,359	1,972
(This work, 2023)	PDC >80% RF		✓		5-fold cross-validation	95.37%	22,596	5,649
(This work, 2023)	PDC = 100% CNN	✓	✓	✓	5-fold cross-validation *(undersampled)*	97.40%	6,338	3,142
(This work, 2023)	PDC = 100% RF	✓	✓		5-fold cross-validation	95.40%	22,596	5,649

When migrating this methodology to testing on the entire patient population (i.e., regardless of how long they have been on service), using 5-fold cross-validation, the AUC increases to 97.4% with a nonadherence prediction precision of 90.1% for the CNN; the RF network attained an AUC of 95.4% and a nonadherent prediction precision of 88.5%. This performance difference signifies that nonadherence prediction is more challenging on patients with lower LOS, this could be explained through different behavior for new patients, as well as the reduction in data quantity. Whilst nonadherence prediction for new patients is our primary focus, performance across the full dataset is still relevant, as it is also important to be able to identify any patients across the cohort who may subsequently become nonadherent and would benefit from ongoing additional support.

When the PDC requirement is lowered to 80%, the AUC of the CNN improves marginally to 97.9% with a nonadherence precision of 87.3%. Whilst the RF network attains the same AUC and a marginal nonadherence precision improvement to 88.6%. These results are more comparable to those found in other studies due to the same PDC requirements being used.

The results of this study, particularly when considered in the context of other similar studies, highlights the need for further research with respect to clearer impact of patient demographics and behavioral patterns on adherence prediction. We acknowledge that the demographics and behaviors of the patients included in all the studies reviewed, and our study, may differ. Similarly, we acknowledge that our study may differ from the other studies that were reviewed, in terms of differing patient diagnosis and their use of a less strict PDC metric.

Our design decision to consider the use of medication stockpiling for our calculation of medication availability is an approach that was rarely seen in other studies and so should be considered when comparing the results. Importantly however, due to the prevalence of medication stockpiling within chronic disease patients, we believe this methodology is more suitable for defining adherence and adds value to this study [29].

IX. CONCLUSION

This study set out to predict nonadherence for patients that are new to a homecare delivery service, where new patients were defined as having been on the service between 3 and 4 months. Our best-performing model achieved an AUC of 82.8% when predicting whether these patients would run out of medication in the next month (by failing to confirm the delivery of their prescribed medication). When adapting this methodology to predict nonadherence for the full cohort of patients within the dataset (using a PDC of 80%, so that more comparable evaluation against other studies could be made), an AUC of 97.9% was achieved. Both CNN and RF networks were evaluated for their capability at this task. The CNN tests were conducted using a novel form of data encoding, producing visually represented medication stock timelines for patients with various additional information encoded within them. This outperformed the RF network by 2% AUC. The results have also shown a performance gain through the inclusion of temporal communication information into the network in addition to medication delivery confirmation data.

Future work by the authors includes additional visualization approaches for the benefit of clinicians, as well as testing these visualizations for adherence prediction. Additionally, more comprehensive experimental testing using a PDC of 80% and cross-validation as well as different strategies for further encoding the temporal delivery communications data within the RF network will be explored. Finally, the use of data oversampling instead of undersampling is a technique that has the potential to further improve performance whilst mitigating class imbalance and would be worth evaluating.

ACKNOWLEDGMENT

This work was conducted as part of a predictive adherence project funded by HealthNet Homecare UK LTD.

REFERENCES

[1] F. Kleinsinger, "The Unmet Challenge of Medication Nonadherence," *The Permanente Journal/Perm J*, vol. 22, pp. 18–033, 2018, doi: 10.7812/TPP/18-033.

[2] Y. Gu *et al.*, "Predicting medication adherence using ensemble learning and deep learning models with large scale healthcare data," *Sci Rep*, pp. 1–13, 2021, doi: 10.1038/s41598-021-98387-w.

[3] R. L. Cutler, F. Fernandez-Llimos, M. Frommer, C. Benrimoj, and V. Garcia-Cardenas, "Economic impact of medication non-adherence by disease groups: A systematic review," *BMJ Open*, vol. 8, no. 1. BMJ Publishing Group, Jan. 01, 2018. doi: 10.1136/bmjopen-2017-016982.

[4] M. Lemstra, C. Nwankwo, Y. Bird, and J. Moraros, "Primary nonadherence to chronic disease medications: a meta-analysis," *Patient Prefer Adherence*, vol. 2018, no. 12, pp. 721–731, 2018, doi: 10.2147/PPA.S161151.

[5] A. Galozy and S. Nowaczyk, "Prediction and pattern analysis of medication refill adherence through electronic health records and dispensation data ☆," 2020, doi: 10.1016/j.yjbinx.2020.100075.

[6] K. Haas, Z. Ben Miled, and M. Mahoui, "Medication Adherence Prediction Through Online Social Forums: A Case Study of Fibromyalgia," *JMIR Med Inform*, vol. 7, no. 2, Apr. 2019, doi: 10.2196/12561.

[7] W. Hsu, J. R. Warren, and P. J. Riddle, "Medication adherence prediction through temporal modelling in cardiovascular disease management," *BMC Med Inform Decis Mak*, vol. 22, no. 1, Dec. 2022, doi: 10.1186/s12911-022-02052-9.

[8] H. Kharrazi, X. Ma, H. Y. Chang, T. M. Richards, and C. Jung, "Comparing the Predictive Effects of Patient Medication Adherence Indices in Electronic Health Record and Claims-Based Risk Stratification Models," *Popul Health Manag*, vol. 24, no. 5, pp. 601–609, 2021, doi: 10.1089/pop.2020.0306.

[9] N. Kim *et al.*, "Predictors of adherence to nicotine replacement therapy: Machine learning evidence that perceived need predicts medication use HHS Public Access," *Drug Alcohol Depend*, vol. 205, p. 107668, 2019, doi: 10.1016/j.drugalcdep.2019.107668.

[10] L. Wang *et al.*, "Applying machine learning models to predict medication nonadherence in crohn's disease maintenance therapy," *Patient Prefer Adherence*, vol. 14, pp. 917–926, 2020, doi: 10.2147/PPA.S253732.

[11] X. W. Wu, H. B. Yang, R. Yuan, E. W. Long, and R. S. Tong, "Predictive models of medication non-adherence risks of patients with T2D based on multiple machine learning algorithms," *BMJ Open Diabetes Res Care*, vol. 8, no. 1, pp. 1–11, 2020, doi: 10.1136/bmjdrc-2019-001055.

[12] H. Kumamaru *et al.*, "Using Previous Medication Adherence to Predict Future Adherence," *Journal of Managed Care & Specialty Pharmacy JMCP November*, vol. 24, no. 11, 2018, Accessed: Sep. 27, 2022. [Online]. Available: www.jmcp.org

[13] B. Akshaya and M. T. Kala, "Convolutional Neural Network Based Image Classification And New Class Detection," in *2020 International Conference on Power, Instrumentation, Control and Computing (PICC)*, 2020, pp. 1–6. doi: 10.1109/PICC51425.2020.9362375.

[14] S. A. Dwivedi, A. Attry, D. Parekh, and K. Singla, "Analysis and forecasting of Time-Series data using S-ARIMA, CNN and LSTM," in *Proceedings - IEEE 2021 International Conference on Computing, Communication, and Intelligent Systems, ICCCIS 2021*, Institute of Electrical and Electronics Engineers Inc., Feb. 2021, pp. 131–136. doi: 10.1109/ICCCIS51004.2021.9397134.

[15] L. Breiman, "Random Forests," *Mach Learn*, vol. 45, no. 1, pp. 5–32, 2001, doi: 10.1023/A:1010933404324.

[16] A. John-Syin, N. Kok-Why, and C. Fang-Fang, "Modeling Time Series Data with Deep Learning: A Review, Analysis, Evaluation and Future Trend," in *2020 8th International Conference on Information Technology and Multimedia : 24 & 25 August 2020*, Selangor, Malaysia: Institute of Electrical and Electronics Engineers Inc., Aug. 2020.

[17] V. Koesmahargyo *et al.*, "Accuracy of machine learning-based prediction of medication adherence in clinical research," *Psychiatry Res*, vol. 294, p. 113558, 2020, doi: 10.1016/j.psychres.2020.113558.

[18] A. A. Semenoglou, E. Spiliotis, and V. Assimakopoulos, "Image-based time series forecasting: A deep

convolutional neural network approach," *Neural Networks*, vol. 157, pp. 39–53, Jan. 2023, doi: 10.1016/j.neunet.2022.10.006.

[19] Y. Zhu *et al.*, "Converting tabular data into images for deep learning with convolutional neural networks," *Sci Rep*, vol. 11, no. 1, Dec. 2021, doi: 10.1038/s41598-021-90923-y.

[20] R. Yarlagadda, V. Kosana, and K. Teeparthi, "Power System State Estimation and Forecasting using CNN based Hybrid Deep Learning Models," in *2021 IEEE International Conference on Technology, Research, and Innovation for Betterment of Society (TRIBES)*, 2021, pp. 1–6. doi: 10.1109/TRIBES52498.2021.9751638.

[21] S. Mehtab, J. Sen, and S. Dasgupta, "Robust Analysis of Stock Price Time Series Using CNN and LSTM-Based Deep Learning Models," in *2020 4th International Conference on Electronics, Communication and Aerospace Technology (ICECA)*, 2020, pp. 1481–1486. doi: 10.1109/ICECA49313.2020.9297652.

[22] D. Prieto-Merino *et al.*, "Estimating proportion of days covered (PDC) using real-world online medicine suppliers' datasets," *J Pharm Policy Pract*, vol. 14, no. 1, Dec. 2021, doi: 10.1186/s40545-021-00385-w.

[23] J. M. Franklin, A. A. Krumme, W. H. Shrank, O. S. Matlin, T. A. Brennan, and N. K. Choudhry, "Predicting adherence trajectory using initial patterns of medication filling," *American Journal of Managed Care*, vol. 21, no. 9, pp. 537–544, 2015.

[24] J. E. Lucas, T. C. Bazemore, C. Alo, P. B. Monahan, and D. Voora, "An electronic health record based model predicts statin adherence, LDL cholesterol, and cardiovascular disease in the United States Military Health System," 2017, doi: 10.1371/journal.pone.0187809.

[25] W. Gao *et al.*, "A Clinical Prediction Model of Medication Adherence in Hypertensive Patients in a Chinese Community Hospital in Beijing," *Am J Hypertens*, vol. 33, no. 11, pp. 1038–1046, Nov. 2020, doi: 10.1093/AJH/HPAA111.

[26] X. Li, H. Xu, M. Li, and D. Zhao, "Using machine learning models to study medication adherence in hypertensive patients based on national stroke screening data," in *2021 IEEE 9th International Conference on Bioinformatics and Computational Biology, ICBCB 2021*, Institute of Electrical and Electronics Engineers Inc., May 2021, pp. 135–139. doi: 10.1109/ICBCB52223.2021.9459205.

[27] C. A. Walsh, C. Cahir, S. Tecklenborg, C. Byrne, M. A. Culbertson, and K. E. Bennett, "The association between medication non-adherence and adverse health outcomes in ageing populations: A systematic review and meta-analysis," *Br J Clin Pharmacol*, vol. 85, no. 11, pp. 2464–2478, Nov. 2019, doi: 10.1111/BCP.14075.

[28] S. Al Zoubi, L. Gharaibeh, H. M. Jaber, and Z. Al-Zoubi, "Household Drug Stockpiling and Panic Buying of Drugs During the COVID-19 Pandemic: A Study From Jordan," *Front Pharmacol*, vol. 12, Dec. 2021, doi: 10.3389/fphar.2021.813405.

[29] E. E. Cameron, S. A. Moss, S. J. Keitaanpaa, and M.-J. A. Bushell, "Pharmacists' experiences of consumer stockpiling: insights from COVID-19," 2021, doi: 10.1002/jppr.1758.

[30] A. N. Sayed, Y. Himeur, and F. Bensaali, "From time-series to 2D images for building occupancy prediction using deep transfer learning," *Eng Appl Artif Intell*, vol. 119, p. 105786, 2023, doi: https://doi.org/10.1016/j.engappai.2022.105786.

[31] Z. Wang and T. Oates, "Imaging Time-Series to Improve Classification and Imputation," Palo Alto, California USA, Jul. 2015. doi: https://doi.org/10.48550/arXiv.1506.00327.

[32] Z. Ahmad, A. Shahid Khan, K. Zen, F. Ahmad, and C. Zeeshan Ahmad, "MS-ADS: Multistage Spectrogram image-based Anomaly Detection System for IoT security," 2023, doi: 10.1002/ett.4810.

[33] X.-Y. Liu, J. Wu, and Z.-H. Zhou, "Exploratory Undersampling for Class-Imbalance Learning," *IEEE Transactions on Systems, Man, and Cybernetics, Part B (Cybernetics)*, vol. 39, no. 2, pp. 539–550, 2009, doi: 10.1109/TSMCB.2008.2007853.

A Review on XR in Home-based Nursing Education

Yan Hu
Department of Computer Science
Blekinge Institute of Technology
Karlskrona, Sweden
e-mail: yan.hu@bth.se

Prashant Goswami
Department of Computer Science
Blekinge Institute of Technology
Karlskrona, Sweden
e-mail: prashant.goswami@bth.se

Veronica Sundstedt
Department of Computer Science
Blekinge Institute of Technology
Karlskrona, Sweden
e-mail: veronica.sundstedt@bth.se

Abstract—**Recent developments using extended reality (XR) technologies have allowed for increased use in healthcare in the last few years. This review paper explores how XR applications are utilized in home-based nursing education, in particular, to identify future challenges and opportunities. The systematic literature review evaluates relevant extracted papers based on publication information, XR technology used for education purposes, target users, and study design and evaluation, including sample size. The results show potential for using XR technologies in home-based nursing education. In particular, Virtual Reality (VR) has become quite popular and the most used to date. However, Augmented Reality (AR) has also emerged as an alternative for the future.**

Index Terms—**Extended Reality; Review; Home-based; Nursing**

I. Introduction

Information and Communication Technology (ICT) has consistently supported higher education with a wide range of applications for decades. ICT has been proven to increase motivation and engagement in studying and enhance the collaboration between educators and students [1]. Due to the Covid-19 pandemic, many universities have shifted (at least partially) their education from campus-based to distance education. Because of this shift, researchers and educators are exploring new ways of applying ICT to support distance education. However, programs such as nursing education which require extensive on-site clinical training, may struggle to adopt ICT into this shift in their curriculum. It requires an appropriate design and application of the technologies. Otherwise, the benefits can be limited.

Home-based healthcare is one such area in which ICT may be adopted into nursing education and practice. According to the World Health Organization (WHO), the percentage of people aged 60 and over will double from 12% in 2015 to 22% in 2025 [2]. In addition, 71% of global deaths are caused by chronic diseases. The increasing aging population has already heavily affected the healthcare systems worldwide. As a result, there is a trend to shift traditional hospital-based healthcare to home-based healthcare [3]. This trend also leads to changes in nursing education, as more home-based nursing education solutions are needed in the future. At the same time, the rapid development of ICT also brings several new opportunities to deliver a better education, even remotely.

Recently, as one of the most impactful ICT, Extended Reality (XR) has been applied in higher education. XR is an umbrella term for all the technologies that add virtual elements to the real-world environment to any extent. It includes Virtual Reality (VR), Augmented Reality (AR) and Mixed Reality (MR), and anything in between [4]. VR is a 3-dimensional, computer-simulated virtual environment that can be explored and interacted with people in 360-degree [4]. AR is a real-time use of digital elements in a real-world environment [5]. MR is an interactive environment that combines a computer-simulated environment and a real-world environment [6]. The relationship among AR, MR, VR, and XR is shown in Fig 1. On the one hand, XR has been applied to education in different ways, with the benefits of improving students' problem-solving skills [7]. On the other hand, XR has also been widely used in healthcare, including disease prevention, treatment, medical training, and education [8]. XR devices are quickly becoming less expensive with the rapid development of technology. Hence nursing education with the help of these devices could play an essential role in increasing engagement, reducing the stress of learning, and creating an immersive experience with cutting cost and time efficiency [9]. This paper aims to explore the current state-of-the-art on applying XR in home-based nursing education and discover its future trends. To achieve the research goal, a systematic literature review will be performed. The research questions of this review are:

- **RQ 1**: How have XR applications been used in home-based nursing education?
- **RQ 2**: What are the future challenges and opportunities to apply XR technologies in home-based nursing education?

Applying XR technology in nursing and medical education is a relatively new concept and has recently gained popularity. Even though fewer works have explored the potential, the technology holds tremendous possibilities in the cross-domain linking XR and remote or onsite medical education. This has been demonstrated successfully with the growing application of VR, AR, and MR in pedagogy [7] [10] in several educational disciplines. This paper can help identify the initial advantages and challenges of leveraging XR technology in home-based nursing education.

II. Related Work

VR has been applied in medical education since early this century. Gallagher et al. proposed that VR could be introduced as a tool for surgical training to improve surgical technical skills in 2005 [11]. Since then, more VR-based

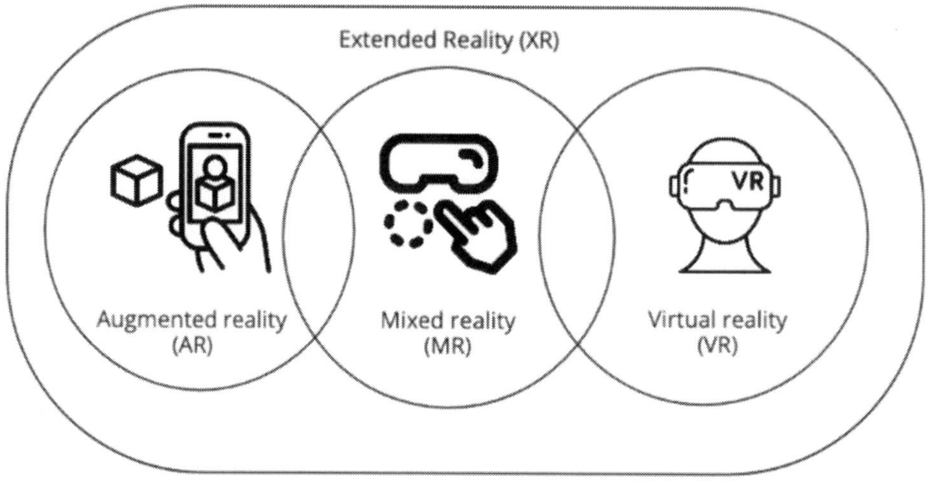

Fig. 1. XR technologies, including augmented reality (AR), mixed reality (MR) and virtual reality (VR).

surgical simulation programs have been implemented and used in different countries. A review of the use of VR in surgery training revealed that four commercially available simulators (dVSS, dV-Trainer, RoSS, SEP) had been demonstrated in doctor and nursing education programs. Most medical students who have used those four simulators have agreed that VR training is an effective way to improve skills and is comparable with the skills training in a lab [12].

Another review of XR game applications in healthcare noted that medical education and training are one of the main aims of applying XR into healthcare [8]. The training or education includes clinical training, basic life support training, nursing skills training, stress inoculation training, etc. [8]. Recently, immersive VR has been reviewed in nursing education [9]. This study indicated that immersive VR improved learning, cognitive, and psychomotor performance. However, the challenges of motion sickness and lack of visual comfort were also reported [9]. Existing review papers already clearly show how XR applications could contribute to nursing education in hospital-based healthcare. Considering the different requirements of home-based healthcare, unlike other reviews listed above, this review focuses on the XR applications in home-based nursing education.

III. METHOD

In order to answer the research questions, a systematic literature review is conducted. A systematic literature review provides a comprehensive and unbiased summary of current literature relevant to research questions. To maximize the coverage of our searched literature, we identified some of the most used words/concepts and synonyms in the research questions. We first conducted a manual search in computer science, healthcare and education. The selected databases were PubMed, Web of Science, ACM Digital Library, IEEE Xplore and Scopus. The publication year was not limited since XR is a recent emerging technology. In addition, the language of the papers was limited to English. The search string listed below was used to search in the selected databases:

- ("virtual reality" OR "VR") AND (nurs*) AND ("educat*" OR "teach*") AND (home)
- ("augmented reality" OR "AR") AND (nurs*) AND ("educat*" OR "teach*") AND (home)
- ("mixed reality" OR "MR") AND (nurs*) AND ("educat*" OR "teach*") AND (home)

The search string is sometimes modified slightly to copy the rules for search strings in different databases. The first search by the search string in all the mentioned databases produced 107 articles. The review processes are mainly based on the guidelines for performing systematic literature reviews in software engineering [13]. First, a primary evaluation was conducted by reading the abstracts and conclusions of all selected articles to explore the most relevant literature to our research questions. Then the second round of full-text review was conducted for the rest of the papers. The criteria applied for the evaluation are described in Table I.

TABLE I
INCLUSION AND EXCLUSION CRITERIA.

Inclusion criteria	Exclusion criteria
Written in English Studies that apply XR in home-based nursing education Have available full-text version Journal articles, conference papers and book chapters	Not written in English Less than 5 pages Review papers

Due to the small number after the two rounds of review, the only quality assessment criteria we applied in this study

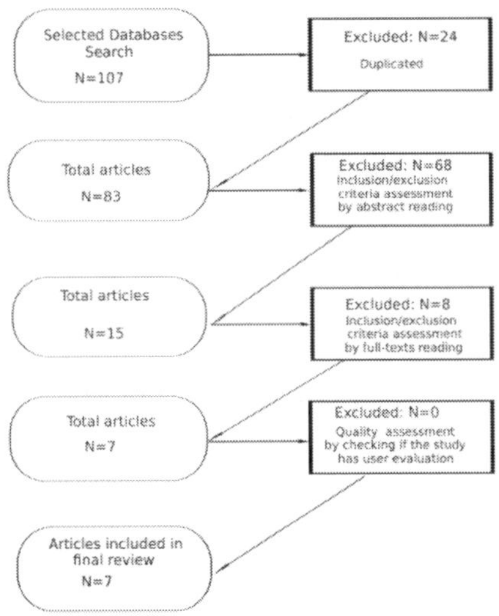

Fig. 2. Flowchart showing processes and results of systematic review

was to check if the paper has a user evaluation. The reason was the user evaluation results could be important inputs to answer our RQ2 compared with the only conceptual designs. After the inclusion/exclusion criteria assessment and quality assessment, data extraction was performed to avoid the bias of subjective preference. A method was applied in which one researcher extracted the papers, and another checked the extraction by reading the abstract of the selected documents. In the end, seven articles were selected for our thorough study, all included in the reference list. The flow chart of the stages and results of the systematic review is shown in Figure 2.

IV. Results

The following sections present the main results found in the review.

A. Publication Information

The selected seven papers were published in the last five years, starting from 2018 to 2022. Prior to 2018, few studies specially applied XR technologies in home-based nursing education. Figure 3 shows the number of papers for each year. The first three years have one paper for each year, but in 2021 and 2022, there were two publications. Among the seven papers, the countries and regions that carried out the studies are the USA (N=2), Spain (N=1), United Kingdom (UK) (N=1), Japan (N=1), Taiwan (N=1), and Brazil (N=1). It is a sign that applying XR in home-based nursing education is considered for broader use worldwide. Six papers applied VR, while one applied AR technology (See Table II). Other detailed information on the selected papers is listed in Table III.

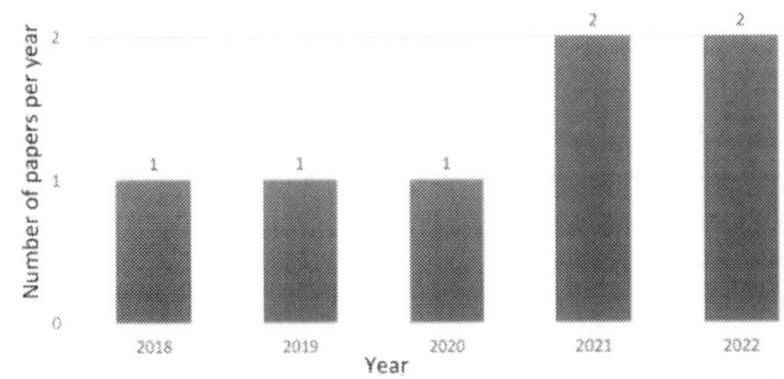

Fig. 3. Number of papers published each year as identified in the review.

TABLE II
XR TECHNOLOGIES APPLIED IN THE STUDIES.

Technology	Number of papers
VR	6
AR	1

B. Education Purposes

All the studies aimed to transfer medical knowledge to the targeted nursing students in their specific areas. Three out of seven studies worked on care-skills training, such as endotracheal suctioning [14], Nasogastric tube care-skill training [15] and Neuroanatomy and Neurorehabilitation [16]. These skills training are not only for home-based healthcare but also apply to hospital-based care. The remaining four studies are more related to home-based healthcare scenarios, with one on general home-based healthcare [17] [18], elderly home-based care [18] and Parkinson's disease and associated disorders care [19].

C. Study Design and Evaluation

The study that applied AR technology [17] designed a mobile AR-based game for nursing students to learn how to deal with different situations when they visit the homes of different groups of people. The combination of mobile and game makes it easier to deliver distance education because of the flexibility and low cost. The designed game could improve the engagement of the students [17]. Most VR applications employ immersive VR, which uses Head-Mounted Devices (HMDs) to run the applications. Some were even assisted with motion capture sensors for interaction with the VR systems [14], [16]. One of the applications, namely Nursing XR, is an integrated platform mainly based on immersive VR technology [20]. The emphasis of the identified seven studies is varied. Some of them highlight the design and implementation of the XR-based education platform [14], [20]. Some others emphasize how the whole XR-based education or training program is designed [16], [18], [19]. In contrast, others are more focused on user evaluation [15], [17].

All seven studies include some user evaluation. Four of them use surveys as the primary evaluation method [16]–[19]. The questions contain standard questionnaires, researchers-self-defined questionnaires, and open-ended questions. The standard questionnaires include the System Usability Scale (SUS) [21], Gameful Experience (GAMEX) [22], QualCare Scale [23], etc. Those questions cover the XR systems' usability and user experience, the system's satisfaction, and the quality of home care. Two studies choose face-to-face communication methods like interviews [15] and workshops [20] to get direct feedback from the users. The study by Komizunai et al. [14] conducted a user experiment as the objective evaluation method. The data collected from the experiment is eye tracking data and motion of the whole body without fingers [14]. All seven studies received positive results on usability and user experiences from the evaluation results. The improvement in learning outcomes is also reported. Notably, the users from some studies pointed out the XR-based applications are fast to learn, free of learning stress, and enchanting the interdisciplinary collaboration [15], [16], [19]. However, the study by Chang and Lai [15] also mentioned that the convenience of the practice needs adaptation.

V. Discussion

Besides the traditional medical skills training, home-based healthcare also needs education on the different case studies in the home environment. It could help the home nurses to be well prepared before they visit different kinds of care recipients. Building up different home environments is unrealistic due to potentially high costs and access limitations. In this case, ICT-based simulations could play a significant role in it.

Previous studies [24], [25] indicated that case-simulation education could improve nursing students' empathy, confidence, and attitudes to dealing with healthcare recipients at home and cooperation in teamwork. The results of this review have similar findings with previous reviews only focused on XR applications in education [26] or healthcare [8], which shows that VR is the most applied technology in home-based nursing education. With its immersive feature, we believe that VR could significantly contribute to the home environment/case simulations for home-based nursing education. 3D virtual anatomical models and 360-degree videos are appropriate options for such simulations.

From our results, we also find the potential for applying AR technology in the future. Unlike VR, AR has its own feature of easy assessment and is not so dependent on auxiliary or other devices. In many cases, a smartphone is enough to run the applications. AR could provide widespread solutions at a low cost for home-based nursing education, which could benefit developing countries. Gamified XR applications for nursing education are another future trend to engage the students' involvement and improve the quality of education. Furthermore, some applications could also be introduced to other people who are involved in home-based care, in addition to the nurses. For instance, to the family members of chronic disease patients or elderly people who live at home, to develop a friendly and supportive community for those care recipients [27].

As with all new applications of ICT, the acceptance of technology in the home-nursing field also needs to be under consideration. Our previous study showed that home-based healthcare professionals could not easily adapt to the new ICT [3]. In addition, VR applications using HMDs can potentially induce motion sickness. This could be a challenge when applying XR technologies in home-based nursing education in the future. The time period of education, the level of integration of the XR technology, and the way of interaction with the systems are all aspects that need to be considered and further discussed with users when applying XR technologies in home-based nursing education.

Since this review only focuses on the XR applications in home-based nursing education, which is quite a narrow scope. The number of the included paper is only seven. It could affect the generalizability of the results of this review.

VI. Conclusions

This study conducted a systematic literature review to discover the current state-of-art XR applications applied in home-based nursing education and their potential and challenges. Seven papers are selected and analyzed based on the country or region, applied technology, education purposes, target users, study design, and evaluation. The results show that VR is the most used technology in this area for an immersive home-environment simulation. It could meet the empathy, confidence, and attitude education requirements for home-based nurses. Additionally, with other sensors, medical skills training could be performed outside the medical labs through VR applications.

AR was also applied in one study, but there is a general trend to develop more applications because of their low cost, flexibility, and accessibility. How to effectively integrate the XR technology smoothly in home-based nursing education and increase the target users' acceptance will be the challenge in the future. It is also essential to evaluate such systems with different target groups, such as patients, relatives, and medical professionals, before use.

Acknowledgment

This research was funded partly by the Knowledge Foundation, Sweden, through the Human-Centered Intelligent Realities (HINTS) Profile Project (contract 20220068).

References

[1] N. Duţă and O. Martínez-Rivera, "Between theory and practice: the importance of ict in higher education as a tool for collaborative learning," *Procedia-Social and Behavioral Sciences*, vol. 180, pp. 1466–1473, 2015.

[2] W. H. Organization. (2022, Oct.) Ageing and health. [Online]. Available: https://www.who.int/news-room/fact-sheets/detail/ageing-and-health

[3] Y. Hu, G. Bai, S. Eriksén, and J. Lundberg, "An iot cloud model for diabetes home-based care: A case study for perceived future feasibility," *Enhanced Telemedicine and e-Health: Advanced IoT Enabled Soft Computing Framework*, pp. 99–115, 2021.

[4] C. Hillmann, *UX for XR: User Experience Design and Strategies for Immersive Technologies*. Springer, 2021.

[5] M. Billinghurst, A. Clark, and G. Lee, "A survey of augmented reality," *Foundations and Trends® in Human–Computer Interaction*, vol. 8, no. 2-3, pp. 73–272, 2015.

[6] M. Speicher, B. D. Hall, and M. Nebeling, "What is mixed reality?" in *Proceedings of the 2019 CHI conference on human factors in computing systems*, 2019, pp. 1–15.

TABLE III
DETAILED INFORMATION OF THE IDENTIFIED PAPERS.

Study	Region	Technology	Educational purpose	Users	Sample size	Evaluation method
Pickering et al. [18]	USA	VR	Elder abuse and neglect	Nurses/social workers	36	Survey
Borges et al. [17]	Brazil	AR	Home visiting	Nursing students	135	Survey
Komizunai et al. [14]	Japan	VR	Endotracheal suctioning	Experienced nurses	12	Experiment
Chang and Lai [15]	Taiwan	VR	Nasogastric tube care skill	Nursing students	60	Interview
Obrero-Gaitan et al. [16]	Spain	VR	Neuroanatomy and Neurorehabilitation	Nursing students	30	Survey
Gilardi et al. [20]	UK	VR	Decision Making	Nursing students/ lecturers	12	Workshop
Hess et al. [19]	USA	VR	Parkinson's disease and related disorders care	Home healthcare professionals	52	Survey

[7] A. Alnagrat, R. C. Ismail, S. Z. S. Idrus, and R. M. A. Alfaqi, "A review of extended reality (xr) technologies in the future of human education: current trend and future opportunity," *Journal of Human Centered Technology*, vol. 1, no. 2, pp. 81–96, 2022.

[8] Y. Fu, Y. Hu, and V. Sundstedt, "A systematic literature review of virtual, augmented, and mixed reality game applications in healthcare," *ACM Trans. Comput. Healthcare*, vol. 3, no. 2, mar 2022.

[9] J. Choi, C. E. Thompson, J. Choi, C. B. Waddill, and S. Choi, "Effectiveness of immersive virtual reality in nursing education: systematic review," *Nurse Educator*, vol. 47, no. 3, pp. E57–E61, 2022.

[10] C. Ziker, B. Truman, and H. Dodds, "Cross reality (xr): Challenges and opportunities across the spectrum," *Innovative learning environments in STEM higher education: Opportunities, challenges, and looking forward*, pp. 55–77, 2021.

[11] A. G. Gallagher, E. M. Ritter, H. Champion, G. Higgins, M. P. Fried, G. Moses, C. D. Smith, and R. M. Satava, "Virtual reality simulation for the operating room: proficiency-based training as a paradigm shift in surgical skills training," *Annals of surgery*, vol. 241, no. 2, p. 364, 2005.

[12] J. D. Bric, D. C. Lumbard, M. J. Frelich, and J. C. Gould, "Current state of virtual reality simulation in robotic surgery training: a review," *Surgical endoscopy*, vol. 30, pp. 2169–2178, 2016.

[13] B. Kitchenham and S. Charters, "Guidelines for performing systematic literature reviews in software engineering," EBSE Technical Report EBSE-2007-01, Tech. Rep., 2007.

[14] S. Komizunai, N. Colley, and A. Konno, "An immersive nursing education system that provides experience of exemplary procedures from first person viewpoint with haptic feedback on wrist," in *2020 IEEE/SICE International Symposium on System Integration (SII)*. IEEE, 2020, pp. 311–316.

[15] Y. M. Chang and C. L. Lai, "Exploring the experiences of nursing students in using immersive virtual reality to learn nursing skills," *Nurse Education Today*, vol. 97, p. 104670, 2021.

[16] E. Obrero-Gaitán, F. A. Nieto-Escamez, N. Zagalaz-Anula, and I. Cortés-Pérez, "An innovative approach for online neuroanatomy and neurorrehabilitation teaching based on 3d virtual anatomical models using leap motion controller during covid-19 pandemic," *Frontiers in Psychology*, vol. 12, p. 590196, 2021.

[17] F. R. Borges, L. C. S. da Costa, C. C. V. Avelino, L. A. de Freitas, C. Kirner, and S. L. T. Goyatá, "Evaluation of an educational technology using augmented reality for home visiting teaching," *Revista Enfermagem UERJ*, vol. 27, p. 37485, 2019.

[18] C. E. Pickering, K. Ridenour, Z. Salaysay, D. Reyes-Gastelum, and S. J. Pierce, "Eati island–a virtual-reality-based elder abuse and neglect educational intervention," *Gerontology & Geriatrics Education*, vol. 39, no. 4, pp. 445–463, 2018.

[19] S. P. Hess, M. Levin, F. Akram, K. Woo, L. Andersen, K. Trenkle, P. Brown, B. Ouyang, and J. E. Fleisher, "The impact and feasibility of a brief, virtual, educational intervention for home healthcare professionals on parkinson's disease and related disorders: pilot study of i see pd home," *BMC Medical Education*, vol. 22, no. 1, p. 506, 2022.

[20] M. Gilardi, S. Honnan, L. Sheerman, A. Cund, S. Rae, and U. Paisley, "Nursing xr–a vr application to teach decision making to student nurses," in *European Conference on Game Based Learning*, 2022, pp. 244–252.

[21] J. Brooke *et al.*, "Sus-a quick and dirty usability scale," *Usability evaluation in industry*, vol. 189, no. 194, pp. 4–7, 1996.

[22] R. Eppmann, M. Bekk, and K. Klein, "Gameful experience in gamification: Construction and validation of a gameful experience scale [gamex]," *Journal of interactive marketing*, vol. 43, no. 1, pp. 98–115, 2018.

[23] L. R. Phillips, E. F. Morrison, and Y. M. Chae, "The qualcare scale: Developing an instrument to measure quality of home care," *International Journal of Nursing Studies*, vol. 27, no. 1, pp. 61–75, 1990.

[24] C. Tran, E. Toth-Pal, S. Ekblad, U. Fors, and H. Salminen, "A virtual patient model for students' interprofessional learning in primary healthcare," *Plos one*, vol. 15, no. 9, p. e0238797, 2020.

[25] W. J. Hwang and J. Lee, "Effectiveness of the infectious disease (covid-19) simulation module program on nursing students: disaster nursing scenarios," *Journal of Korean Academy of Nursing*, vol. 51, no. 6, pp. 648–660, 2021.

[26] L. Coyne, T. A. Merritt, B. L. Parmentier, R. A. Sharpton, and J. K. Takemoto, "The past, present, and future of virtual reality in pharmacy education," *American journal of pharmaceutical education*, vol. 83, no. 3, pp. 281–290, 2019.

[27] D. W. Sari, A. Igarashi, M. Takaoka, R. Yamahana, M. Noguchi-Watanabe, C. Teramoto, and N. Yamamoto-Mitani, "Virtual reality program to develop dementia-friendly communities in japan," *Australasian Journal on Ageing*, vol. 39, no. 3, pp. e352–e359, 2020.

Promotion of Wellbeing in Japanese Culture using Positive Computing

Isabel Schwaninger
Digital Medicine Group
LCSB, University of Luxembourg
Esch-sur-Alzette, Luxembourg
isabel.schwaninger@uni.lu

Sissi Zhan
Human-Computer Interaction Group
Faculty of Informatics, TU Wien
Wien, Austria
e1325880@student.tuwien.ac.at

***Abstract*—Many Japanese tend to associate mental health issues with a weak personality, which leads to people hiding their problems and avoiding help-seeking. Thus, this paper presents user research and design of a smartphone application tailored to people with a Japanese background that promotes wellbeing using positive psychology, focusing on positive feelings and events as well as methods to increase positive emotions. To this end, a narrative literature review has been conducted on mental health in the Japanese context and positive computing/positive psychology, which was followed by five qualitative interviews with participants with a Japanese background living in Europe. The literature review and interviews resulted in 24 design implications (12 for each step), and a high-fidelity prototype for an application to foster mental wellbeing through self-help.**

Keywords- ***wellbeing; Japanese culture; positive computing; mental health app; qualitative user research***

I. Introduction

Mental health tends to viewed as a taboo topic by the general public in Japan, where people tend to believe that mental health concerns are the effect of a weak personality instead of a real health issue [1]. Showing mental health symptoms is also highly associated with feelings of shame due to concerns to be perceived as weak by other people or the community in general [2]. The consequence of this concern is that people tend to avoid seeking help from friends, family or professionals and prefer to handle their problems on their own [3]. While much psychological research applied in Human-Computer Interaction (HCI) is tailored to Western contexts, people with other cultural backgrounds also live in Western contexts like Europe, where they often use technology designed with Western values. Therefore, taking views of users with Japanese background into account when designing digital interventions for mental wellbeing could be beneficial.

The aim of this paper is to present a concept for a smartphone application that promotes mental wellbeing for people with a Japanese background living in Western contexts with the idea to focus on self-help and interventions to be carried out alone for people who avoid help seeking. For this, suitable practices from positive psychology [4] were integrated in the conceptual application, in particular gratitude and humor interventions. Furthermore, qualitative interviews with five participants with a Japanese background living in Europe were conducted to specify the concept and requirements in detail.

This paper is structured as follows. In Section II, we present the approach and methods used, including a narrative literature review and qualitative interviews to gather data, and a short description prototype implementation procedure. Results of the literature analysis are presented in Section III, followed by the interview results in Section IV. The final design requirements and the concept are then presented in Section V. Section VI discusses the outcome and limitations of the research, followed by the concluding Section VII.

II. Methods

The identification of requirements and implementation of the prototype was conducted in three phases: (1) narrative literature review, (2) qualitative interviews, and (3) concept and prototype design. From the literature review, we pulled out a set of design implications (see Table I). These formed the basis for defining questions for the qualitative interviews (see Table II). Extensions and adjustments were made to the design implications after gathering insights from the interviews. Based on these results, a high-fidelity prototype of a smartphone application was created.

A. Phase 1: Literature Review

Research papers for following areas were relevant for this work: perception of mental health in Japanese society (including stigmatization and causes of stigma), positive psychology & positive computing (including a general outline and its focus on Western contexts, gratitude interventions, humor interventions), and user engagement and motivation (including the use of chatbots & avatars for promoting wellbeing). Several journals and conference proceedings were screened for searching papers: Human-Agent Interaction (HAI), Human-Computer Interaction (HCI), PubMed, and JStage specifically for Japanese papers. Keywords used in literature search were combinations of the following: Japan, mental health, perception, stigma, mental wellbeing, cultural, positive psychology, positive computing, intervention, gratitude, humor, chatbot, woebot, avatar, gatebox as well as variations of the above words. In total, 43 papers were reviewed. The papers and topics were clustered and analysed using the *Miro* software.

B. Phase 2: Qualitative Interviews

Five qualitative interviews were conducted with students with a Japanese background living in Europe, who were all acquired through a common acquaintance. These interviews were hold as online video conferences through Jitsi or Zoom using English for communication. The format was a semi-structured interview using an interview guide containing questions that addressed how participants personally deal with negative emotions, their perception of gratitude and humor interventions and their impressions of digital mental health companions, in particular the application Woebot and avatars for chatbots, in particular the device Gatebox. Furthermore, presentation slides were used during the interview as an addition to the questions for showing pictures and videos. The interviews were video-recorded and semi-transcribed as interview notes in the next step. Reflexive thematic analysis [5] was conducted to identify themes across the data.

C. Phase 3: Prototype Implementation

For the concept, a high-fidelity prototype was created to depict the resulting design implications. First, the prototyping tool *Figma* was used to build the design of different smartphone application screens, which were then exported to the prototyping tool *Protopie* for implementing interactivity and logic functionality allowing the prototype to react to user input.

III. Literature Review

To inform requirements from previous research, a narrative literature review was conducted. The results will be presented in the following, covering the topics *A. Perception of Mental Health in Japan*, *B. Positive Computing*, and *C. User Engagement and Motivation.*

A. Perception of Mental Health in Japan

1) Stigmatization: Mental health topics such as depression, schizophrenia and suicide are still heavily stigmatized in Japan and seen as a taboo [1]. Moreover, there tends to be a lack of awareness regarding mental health in general, as many Japanese (1) do not believe in the effectiveness of treatments or in a high chance of recovering, or/and (2) blame the weakness of one's personality or lack of willpower as the cause of mental health issues [1], [2], [6].

This mindset leads to the problem that people who suffer from mental health issues often feel shame and self-criticism for their concern and do not seek help from family, friends or professionals [2], [7]. People tend to avoid talking to other people about their problems due to the fear to be perceived as weak by others, which can result in feeling the need for social isolation [3], [8]. Additionally, acquaintances of people with mental illnesses even tend to socially distance themselves from a person who is affected by a mental condition, especially those with a closer relationship [1].

2) Cause of Stigma: Possible factors that have been studied to potentially have an influence on stigma towards mental health conditions. There are individual factors, such as socio-demographic data like age and gender, as well as socio-economic data such as education and occupation. Moreover, knowledge and familiarity of mental illnesses could also be influencing factors, as well as if a person knows or has been in contact with other people with mental illnesses [9]. Ando et al. [1] have also stressed that missing knowledge contributes to stigma. Apart from this, there is a lack of provided education and national programs in Japan to help tackle incorrect knowledge about mental health issues and provide insights from latest research. Then, there are communal factors, such as the social capita which is, e.g., composed of social networks, trust or relation between people that create a shared believe and social rules among the community [9]. Some also see cultural factors as a cause of negative mindset towards mental problems and emphasize that the norms in Japanese culture and community lead to such attitudes [2], [10]. In general, mental health issues tend to be seen as a topic to avoid in discussions, and individual feelings tend to be disregarded or hidden [1], [11]. A study further showed that people were aware of the presence of their mental health issues, however lacked knowledge about solutions for their problem and countermeasures [11]. Research further suggests that people in Japan tend to feel more comfortable with implicit support, contrary to the preference for explicit support in Western contexts [12].

B. Positive Computing

1) Fostering Wellbeing in Western Contexts: In recent years, the idea to promote happiness and well-being of humans through computer-based technologies grew increasingly in the HCI community. This led to the emergence of the area of positive technology or positive computing, which is a interdisciplinary field of two studies: HCI, which concerns itself with the incorporation of a human-centered approach in the design process of ICT applications and positive psychology [13], [14]. The study of positive psychology aims to identify and understand positive emotions, positive character and finding ways to increase them, as well as the institutions that enable their flourishing [15]. Thus, the design of strategies or so-called interventions to foster positive feelings is one of its key aspects [16]. As a result, positive technology consists of a theoretical pillar, that develops concepts and frameworks to integrate positive psychology into technology and a methodological part for the design, implementation and evaluation of applications to enhance positive affect [14]. To foster wellbeing in and through technology, determinant factors and strategies for improvement can be used, which showed a direct impact on humans wellbeing. Higher level of these factors also result in higher levels of wellbeing. Some identified examples for determinant factors are positive emotions, self-awareness, mindfulness, gratitude, compassion [4].

However, positive psychology is strongly influenced by Western or more precisely, North American mentality, which has been criticized often [17]. As Western countries developed a more individualistic cultural mindset, researched interventions and strategies tend to focus on improving individual wellbeing, which are not ideal to implement and may not be as effective for users with collective background that sets great value on the community and how the individuals can contribute to the whole [18].

Due to the above stated cultural influences in the state-of-the-art interventions of positive psychology and the importance of incorporating cultural aspects when developing applications to foster wellbeing, our approach will take characteristics of Japanese culture and perceptions of users with a Japanese background on mental health conditions into consideration. We argue that even for people with Japanese background living in in Western contexts, this could be beneficial. Target users tend to avoid talking about personal problems and seeking help from friends, families or professionals, and they may even feel guilt for receiving social support [12]. Therefore, our approach is to focus on interventions and strategies not based on social factors but instead on self-help, self-care or anonymous interventions.

2) Interventions using Gratitude: Gratitude has been listed as one of the top character strengths that have been robustly and consistently related to life satisfaction [19]. Thus, interventions focusing on gratitude have been developed and seem to be promising for increasing the level of wellbeing and happiness as well as for reducing negative feelings [20], [21]. In long-term studies, it has been reported that practicing gratitude gives good results in improving mood, such as in the work of Seligman et al., in which their participants were asked to complete the three good things intervention over a duration of six months. This intervention requires to think of three good things at the end of every day that occurred on that day and write them down. They were also asked to describe the cause of the good thing. After one month positive effects could be seen: participants reported higher levels of happiness and lower levels of negative feelings [15].

However, multiple studies investigated gratitude practices in a cultural context, partly comparing two groups of different cultural backgrounds and reported that participants from East Asia are not as positively affected from the interventions as those from Western cultures due to their collectivist culture [22]–[24]. Although it can lead to a decrease in negative emotions, gratitude had no significant influence on positive emotions and wellbeing [22]. Though, one common aspect of the named papers is that they defined gratitude in relation to other people, e.g., being thankful for others or for the action of others. Though, there are also studies that do indicate the effectiveness of gratitude exercises for Japanese users, such as one conducted by Otsuka et al., which reported that practising gratitude is beneficial for increasing positive affect and happiness among Japanese workers [25].

Although there are several studies that show a low compability of gratitude interventions to improve wellbeing in collectivistic societies, such as Japan, this work will still incorporate gratitude strategies. As highlighted, often times gratitude only refers to showing thankfulness to other person or even directly communicating one's thoughts of gratitude to a recipient (gratitude visit), which could also cause negative feelings. Nevertheless, gratitude can also be defined much broader, e.g., as gaining awareness of the good things in one's life, being thankful for them and taking time to show thankfulness [19].

3) Interventions using Humor: The idea to put the character strength humor into use to promote wellbeing and the relation of these two factors have been explored in multiple studies, e.g., [26], [27] and others which were named by Ruch and McGhee [28]. The respective studies showed evidence in favor of using humor interventions, as a high level of humor seems to be in accordance with higher levels of wellbeing and less negative emotions [28]. According to [29], there are different ways to attain happiness with one them being based on hedonism, which means to increase the amount of pleasure while decreasing the amount of pain as much as possible. The trait with the highest correlation to pleasure in life has been shown to be humor [29].

One example of a specific intervention using humor is the adaption of other interventions, e.g listing three funny things instead of the traditional three good things, as seen in Gander et al.'s work, which additionally used the adapted intervention in an online format [26]. The participants were instructed to think of three funny things that happened for each day and note them. In contrast to a placebo control group, the humor intervention group showed an increase in happiness as well as a decrease in negative feelings. In another study [27], the following intervention strategies showed to be effective to increase feelings of happiness: (1) counting the amount of funny things happened on the day and writing down the total number, (2) applying humor to daily life meaning to be more aware to humorous experiences and notice them but also incorporating humorous activities, such as reading jokes, enjoying comedy movies or books (3) and the before mentioned three funny things a day intervention. Tsukawaki et al. [30] studied the relation between wellbeing and different types of humor. Both adults and children who belong to the type with a self-defeating humor interestingly showed the highest level of wellbeing compared to the other humor types playful and aggressive humor. Further, Tsujita and Rekimoto investigated how forced smiles using technology could improve the mood of participants, namely through smile therapy. The effects of technology-induced smiling showed positive effects on feelings of happiness when tested on a small participant group [31].

C. User Engagement and Motivation

One concern of digital psychology interventions is that an online intervention can only be effective when it is used by users. Therefore, one important aspect is to achieve user commitment and creating enjoyment, which raises the questions how to (1) make users choose an application, (2) get users to engage with it, and (3) motivate users to keep using the application [32]. Equipping the online intervention with a chatbot and an avatar may present potential methods to raise user experience.

1) Chatbots: Chatbots are used in many sectors and have also found its way into psychology. One of them is a smartphone application called Woebot, which uses a conversational agent combined with researched psychology techniques to help people to overcome negative thoughts and feelings, deal with stress and anxiety and improve their mood [33]. This is done by checking in with the user on a daily basis, and recommending different interventions based on their personal problems. Another aim of the application is to build rapport with users through a human-like technology that shows empathy and thus improving the efficacy of the interventions, which has been noted as a factor that other digital solutions to improve mental wellbeing [34]. Positive reviews in the app store reported that, e.g., using it felt very personal, helped reduce their anxiety, but also that users felt support when they couldn't talk with other people about their problems and emotions and did not feel judged about their feelings by the chatbot [35], [36].

2) Avatars: In recent years, the exploration of avatars in mental health strategies has been increasing and lead to promising results when used, e.g., in interventions for improving mental wellbeing, reducing anxiety, and lessen depressive symptoms [37]. There are multiple ways to integrate an avatar, such as for representation of the client in online therapies in a virtual space, the support of interventions in face-to-face therapy or as autonomous virtual therapists. The latter one describes a virtual agent playing the role of a virtual health coach with a graphic representation, i.e., an avatar to embody the coach [38]. Studies that used such avatars reported great efficacy, e.g., for imparting knowledge and promoting self-care [37]. In addition, when technologies combine an avatar with a conversational agent and thus providing a talking avatar, they have a higher potential to build a prolonged relationship with its users and hold a greater appeal and persuasiveness compared to technologies without such components [37].

Some studies also investigated the influence of different types of avatars and suggested that avatars with a similar appearance to the users are more persuasive compared to arbitrary ones [37]. The possibility of personal customization of avatars can additionally increase user enjoyment and support longer lasting engagement with the application [32].

Another way avatars can be differentiated is between realistic representations depicting real people and animated characters. [38]. One example of an animated avatar is the Japanese Gatebox, a virtual holographic-like character that can answer questions and be used as a smart home device [39]. The Gatebox can be compared to other home assistants such as Alexa, Google Assistant and Co, however its aim is to provide emotional support and act more as an virtual companion than an assistant, while imitating a female voice [39], [40].

IV. Qualitative Interviews

To further specify the requirements, five qualitative interviews were conducted. The age of the participants were between 20 and 30, and all the five participants grew up in Japan and have been also living in Europe for several years, e.g., for studies or work. Furthermore, as far as it is known, none of the recruited people have a case of psychiatric diagnosis and are thus cognitively able.

The interviews lasted around 30 minutes, and questions addressed how participants deal with negative emotions, help-seeking, gratitude and humor interventions, the application Woebot and chatbots in general, the product Gatebox and avatars in general. Participants were asked to try out the Woebot application before the interviews were conducted. The results will be presented in the following.

A. Dealing with Stress and Negative Emotions

Participants used a variety of activities to deal with negative emotions, such as listening to music, watching movies or concentrating on work, studies, hobbies or any other activities. One participant also likes to visit their neighbourhood cat to cuddle it to reduce their stress. Furthermore, sleep and alcohol were also listed to be helpful. Thinking positively was also applied, such as one participant explained: "I just tell myself repeatedly 'it will be alright, it will be alright', it's really simple". Another participant also showed a positive mindset, as they said that what happened can not be changed anymore so it is best to try to forget about it and move on.

B. Help-Seeking

Several participants stated they usually avoid asking for help from other people due to them not wanting to bother anybody and having the mindset to need to solve their problems on their own. Moreover, they want to talk about nicer things and do not want to reveal their 'bad side' to others. However, consulting friends and family was also often mentioned as a method participants usually use when they encounter problems. They stated they feel better after talking, can organize their thoughts better and the cause of their stress gets clearer. They also encourage other friends to talk to them if they have problems. One important aspect that has been stated when providing support to other people is to listen to their feelings and to 'accept what they are'. Furthermore, it was emphasized to not provide advice unless the person specifically asks for it. Participants who seek help from other people find it helpful as talking to other people allows them to organize their thoughts and understand their situation or cause of their stress better.

C. Journal Interventions

For the most part, participants could imagine practicing journaling exercises, such as three good things, gratitude, and three funny things as an intervention, and they had good attitudes towards theses interventions. Participants commented that looking back on the support they received made then feel happy or that these exercises could help them focus on good aspects. However, many thought the integration into daily life and forming a habit to write a journal entry everyday would be difficult. In addition, they worried about not being able to recount everything at the end of the day but on the other hand writing down good or funny events every time something happens is not realistic either.

D. Applying Humor

Doing funny activities, such as watching funny movies or reading funny books and comics were in general well accepted by the participants. Nevertheless, there were also concerns that these exercises could have a counterproductive effect, as people may feel they have wasted their time instead of working something important and thus feel more stressed. In addition, the idea of a daily smile camera has been commented as weird and creepy by some participants.

E. General Attitudes towards Chatbots

Regarding the idea to use chatbots as an emotional companion in general, participants had various different views. Although participants believed that writing down their throughts and feelings is helpful for moving forward, they would also not consider using such a chatbot as the first choice but rather as an alternative when they have no one to talk to or only when they feel stressed and get a notification. In that case it may be nice to interact with a chatbot to talk about what is bothering them or get sent some funny content to uplift their mood: 'maybe the chatbot could even recommend some funny videos or photos of cute animals or maybe even flowers to make users feel more relaxed and affect their mind positively or it could propose some activities in order to manage what happened, like keep me talking and thinking about what happened'.

On the other hand, participants also perceived chatbots only as a service that provides information and they wished to interact with an 'objective chatbot', which only replies with information about exercises and tips for shifting the mood and also shows evidence for why they are effective. This was especially mentioned for 'normal times' when participants would not feel negative emotions or stress: 'For normal times I would need some objective information and platform, they can just provide me with information for how to deal with the problem better and tell me it's good to do this exercises'. Another topic that participants would find interesting is seeing some analysis of their own data about mood, stress and emotions. They talked about the chatbot being able to recognize when users are stressed and that users can monitor their own mood patterns and being shown tips to improve their situation.

F. Chatbot Personality

Regarding the personality of chatbots, many participants talked about a calm personality that is friendly and shows empathy. Especially for difficult times, the chatbot should be 'empathic, show understanding and encourage the user', however it should not be 'too emotional' either. It seemed important for the participants that the chatbot listens to their problem and support them to keep talking instead of acting too 'persuasive or aggressive' to provide solutions and answers. On the other hand, some participants also wished for a solution-focused chatbot that can give good advice. It was also mentioned that the chatbot 'should not be too human-like and rather objective to some extent, just for exercise clearly and for tasks or activities'.

G. Attitudes towards Woebot

For the specific chatbot Woebot, the participants did not try out the app for longer than a couple hours, and some did not download it at all. Those who did use the application perceived it as repetitive, as 'it keeps asking the same questions' and suggests users interventions but does not explain them beforehand and thus received as troublesome.

H. Attitudes towards Gatebox

When the participants were asked about their first impression of the gatebox device, many perceived the device as 'weird', 'strange' or even 'creepy', as one person pointed out:'It's creepy because it acts like your girlfriend'. It was also mentioned that they see gatebox as 'nerdy stuff' and something that is targeted towards 'otaku culture', hence it was also often stressed by participants that they 'personally don't need it'. Nonetheless, the general idea of having a virtual character as a companion was accepted and participants could see that for some people it can be very helpful if they feel lonely or have no one to talk to. In that case, the avatar should appear real, as stated in some interviews: 'if they say good morning every day I will get bored and think it is a machine' or 'I would think it is a robot and wouldn't believe its emotional affect, I would think they are not really thinking about me and understanding me'.

I. General attitudes towards Avatars

In the interviews different kinds of avatars were discussed and compared: real person, animated person, non-human, media characters. All participants did not like real human pictures as avatars, as explained in the interviews: 'Real people would be weird, I would feel like who are you', 'I would definitely not like realistic ones, I don't want to talk with them' or 'The realistic avatars are maybe too creepy, it's like a real person in my home'. One participant also expressed that it would feel easier interacting with non-human avatars: 'If it is a real or animated person I would think they're too real, I wanna talk to someone who is not thinking anything, maybe it would be easier to talk because I think they would forget about it next day'. The majority of participants preferred either non-human avatars such as cute animals or some media characters they are familiar with, examples mentioned were, e.g., characters from marvel, animal crossing, favourite actors, mascots from their favourite groups. One participant specifically mentioned 'I can feel more familiar with media characters, with other characters I don't know them. With media characters I can feel or imagine that we are friends'.

1) Frequency: The frequency of engaging in interventions for wellbeing and the willingness to be notified by the application varied greatly from person to person, from daily interaction to notifications only when feelings of stress occur. Some participants noted that it would be bothersome if they get notified regularly and would only use the application when they are in a bad mood. In particular, one participant mentioned that ideally they would only receive a notification when the application recognizes a high stress level, e.g., through analysis of users mood pattern.

V. Concept

The concept aims to promote mental wellbeing for users with Japanese background, focusing on self-help and selfcare without being dependent on social relationships in order to avoid feelings of bothering others and indebtedness caused by cultural aspects and characteristics of collectivist societies. The concept of a smartphone app has been created based on the literature review and user research. The ideas have been collected into the concept for a smartphone application that integrates psychological interventions.

A. Psychological Interventions

In the literature review, the two wellbeing factors gratitude and humor to increase wellbeing were explored.

Regarding gratitude, the particular intervention named was the 'three good things' practice, which consists of thinking about three good things that happened that day and writing it down at the end of every day. Good results were reported for improving positive affect and reducing negative feelings. Therefore, the application will prompt users to perform the three good things practice on a daily basis. However, the definition of gratitude will not be limited to gratitude towards other people or to the actions of other people (e.g., 'I'm thankful my friend helped me with a problem') and will take on a broader sense: possible gratitude entries are e.g., 'I'm grateful that I could take a walk in the park today', or 'I'm thankful for this movie which I enjoyed'.

One intervention based on the factor humor is a variation of the 'three good things' practice, which has been modified to the 'three funny things' practice. Other effective humor practices are counting the amount of funny things each day, applying humor to daily life and solving stressful events with humor. Additionally, smile therapies have also shown potential to uplift mood and wellbeing, including technology-induced smiling.

Therefore, the application will provide a toolbox of different interventions and prompt users to carry them out regularly. For this, explicit design implications have been derived out of the results of literature analysis and the conducted interviews. Table I and Table II describe the resulting requirements for the application concept based on literature and interviews respectively, the reason for including the requirement and on which section it is referring to.

B. Emotional Support Companion

Similar to the Woebot, an intelligent chatbot will be integrated to guide users through interventions. As target users tend to avoid to seek help from friends and family and prefer to keep problems to themselves, the chatbot should moreover show empathy and embody a companion for emotional support to which users can without social barriers.

In addition, the chatbot will be represented by an avatar to support building rapport with its users, which (1) increases their trust and thus willingness to talk about their problems, and (2), to improve user enjoyment to prolong their engagement with the application and included interventions.

C. Prototype Walkthrough

1) Starting the Application: At the beginning the application displays a starting screen with a character from the game Animal Crossing as an animated non-human avatar. The user is asked how frequent they like to be notified by the avatar or the application, which covers requirement IA11.

2) Home: The homescreen allows the user to access all functionalities of the application: opening the personal journal of the user, saved activities, exploring new or other journal or activity exercises that users can try out. It also depicts the avatar with a message, to which the user can answer. Two buttons at the top of the screen are for recording the current mood and the smile camera to take a photo of the user smiling. Alternatively, a navigation bar at the bottom also leads to the journal, activities, explore page but also to the mood tracker, which shows all recorded mood data of the user.

3) Journal: A daily journal entry can be created for gratitude and humor interventions that involve writing notes such as a gratitude journal or count the funny things exercises, described by requirements LA04, LA05 and LA06 (see Figure 1). However, there is no predefined schedule for any of the interventions, thus letting users fully decide the time and frequency of exercises, which refers to IA01 and IA02 of design implications. Moreover, the journal is flexible and can be extended with preferred journal exercises to make it possible for users to configure their journal entry each time. For example, they can choose to write down three good things only or they can add three funny things and a gratitude journal as well if they feel like it. After saving, changes in the users' mood are tracked.

4) Activities: The activities screen lists all activities saved by the user, i.e. exercises to integrate humor into daily life such as reading a funny book or watching a funny movie. A detailed explanation about these interventions can be displayed, and users can then conduct the exercise according to the given instructions and mark if they have completed an activity, depicting implication LA06. Again, the user is asked about mood changes to track the effectiveness of the intervention.

5) More Exercises: In the explored section, users can slide through different exercises they haven't saved yet and click on one to learn more about the exercise, its effectiveness and instructions. Based on requirements LA03, IA04 and IA06 the application also offers the possibility to learn more about it through a more detailed explanation or by reading through the sources attached in the app. Using the save button, users can add the intervention to their favourites.

6) Chatbot: Users can talk with the chatbot or the avatar if they feel like it. The purpose of the chatbot is to play the role of a virtual friend that users can seek help from at any time. The chatbot reacts to messages of the user and answers in a friendly manner and shows empathy when users talk about their emotions, thoughts and worries. Furthermore,

TABLE I
DESIGN IMPLICATIONS FROM LITERATURE REVIEW

ID	Reason	Section	Implication
LA01	Japanese tend to keep their problems to themselves and avoid talking to other people	Perception of Mental Health in Japan	The app focuses on self-help: interpersonal aspects should not be a requirement for performing interventions (interpersonal interventions are, e.g., thanking another person, social network elements, gratitude intervention focused on other people)
LA02	Lack of education about mental health issues contribute to stigma	Stigmatization	Offer educational content about mental health
LA03	Some people are aware of mental health issues but lack knowledge about solutions and countermeasures	Cultural Comparison	Offer educational content about psychological interventions
LA04	Gratitude interventions may increase positive feelings	Gratitude Interventions	Integrate gratitude interventions as exercises, such as the following: • three good things • up to five things I'm grateful for
LA05	Interpersonal gratitude could cause negative feelings, such as feelings of indebtness and guilt for Japanese	Gratitude Interventions	Do not limit gratitude interventions to interpersonal gratitude
LA06	Humor interventions may increase positive feelings	Humor Interventions	Integrate humor interventions as exercises, such as the following: • three funny things • count the funny things • applying humor to daily life • solving stressful situations with humor
LA07	Taking a picture of oneself with a smile may increase positive feelings	Humor Interventions	Integrate a 'smile camera', which users can use to take a picture of themselves smiling
LA08	A chatbot may improve user engagement	User Engagement and Motivation	Integrate a chatbot, that users can talk with
LA09	A chatbot with empathy may improve user rapport and thus efficacy of interventions	User Engagement and Motivation	The chatbot should act empathic
LA10	Conversational avatars may improve prolonged user relationship and persuasiveness. Avatars may support imparting knowledge and promoting self-care	User Engagement and Motivation	The chatbot should be represented through an avatar (graphic representation)
LA11	Avatars with a smiliar appearance to the users may be more persuasive	User Engagement and Motivation	The chatbot should have a Japanese appearance
LA12	The possibility to customize avatars may improve user enjoyment and engagement	User Engagement and Motivation	Provide the possibility to customize avatars

it aims to uplift their mood through encouraging words, showing understanding, and by sending funny content, such as humorous pictures of animals. Another purpose is to build rapport with the user and improve user engagement to promote their selfhelp. This covers requirements LA08, LA09, IA05, IA07 and IA08.

7) Mood Button: To track the current mood without doing exercises, the mood button can be clicked, which opens a simple pop-up with the avatar asking how the user is feeling.

8) Smile Camera: Requirement LA07 describes possible effects on wellbeing when smiling at a camera. For this, the camera icon navigates to the smile camera, which instructs users to take a picture of themselves while smiling. A picture of a happy smiley is overlaid with the camera to remind users to show a smile.

9) Explore & Learn: In the explore and learn page, users are provided with learning material about mental health and various interventions to improve wellbeing categorised in journal and activities interventions. The icons either open learning content about a specific intervention or topic in the application or links to external material such as blog articles or online videos. Thus, users are offered educational content, as defined in LA02, LA03, IA04 and IA06.

10) Mood Tracker: Finally, the mood tracker screen displays a calendar that shows all mood entries made by the user (see Figure 2). For each day with an entry it is possible to show the tracked mood for the given day and the specific exercises done that day. Based on this information, the application also suggests the top exercises with the highest effectiveness specific for that user. This feature refers to IA12 from the design implications.

VI. DISCUSSION

The aim of this paper is to explore positive psychology to promote wellbeing and self-help for Japanese people living in Europe. While previous research indicates stigmatization

TABLE II
DESIGN IMPLICATIONS FROM INTERVIEWS

ID	Reason	Section	Implication
IA01	It may be a challenge for users to form a habit for daily journal exercises	Journal Interventions	Users should be able to choose the frequency of journal interventions: daily, every second day, every third day or every week.
IA02	Interventions, especially those with longer duration such as watching a funny movie may be perceived as time-wasting for users	Applying Humor	Interventions should not be scheduled as a regular activity
IA03	Participants did not like Woebot's repetitive questions and	Woebot	The chatbot should have variety in its dialog
IA04	Participants wished for more explanation before trying out an intervention in Woebot	Woebot	The application should provide information and explanations about interventions before suggesting them to users
IA05	Participants wish to be sent funny content to uplift their mood	Chatbots	The chatbot should not only talk with users, but also show them funny and uplifting content,e .g. funny pictures/videos, pictures of cute animals, pictures of flowers
IA06	Participants wish for a solely informative chatbot for normal' times when they are not in a bad mood	Chatbots	User should have the possibility to ask the chatbot for information about exercises, tips for uplifting mood and scientific evidence.
IA07	Participants wish for a calm and friendly chatbot personality	Personality	The dialog of the chatbot should appear calm and friendly.
IA08	Participants wish for a chatbot that shows empathy, understanding and encourages the user	Personality	The chatbot should show empathy and reply with encouraging answers
IA09	Some participants only want to be listenend to while others want to receive advice.	Personality	The chatbot should first ask for user's preference to give advice
IA10	Participants preferred non-human avatars or media characters	Avatar	The chatbot should be represented by a non-human avatar or media character
IA11	Participants may feel bothered by frequent notifications	Frequency	The application should ask users about their preferred frequency of notifications: multiple times a day, daily, every couple days, weekly
IA12	Participants wish to see analysis of their mood, stress and emotions	Chatbots	The application should provide analysis for a user's mood, stress and emotions

of mental health in Japan [1], [2], [6], HCI research and wellbeing tends to be focused on users from Western contexts.

The prototype presented in this paper is a result of both design implications from a narrative literature review and views expressed by interviewees with Japanese background, implementing gratitude and humour interventions, and in addition, implementing a chatbot to provide companionship and engagement.

Through the interviews, it could be seen that participants needs were very diverse (also found previously with other user groups, e.g., [41]), in particular, regarding the idea of a chatbot that embodies a virtual companion. Some interviewees found the chatbot aspect unnecessary and preferred a solely informative application to access knowledge about different positive psychology interventions without the need to communicate with a chatbot. Other participants showed a high interest in a virtual companion which they can share their thoughts, feelings and problems with. As also proposed for the design of other devices and use cases (e.g., [42], [43]), this indicates the need for a personalised application to promote wellbeing.

Some participants initially rejected the idea of an avatar stated that they would perhaps think differently about it when they actually feel stressed or other negative emotions. It can be difficult to imagine how they would feel in different moods. Therefore, flexibility of the application is certainly needed.

It was also mentioned that notifications should not appear frequently, but only when needed, e.g., when the user feels stressed. Ideally, the application would be able to recognize mood changes of the user in real-time and only then send a notification. This would require additional solutions, for example, incorporating the use of wearables that are able to identify bad mood through biosignals [44]. Another approach could be the use of artificial intelligence to recognize patterns or seasons in mood data.

During the interviews, the device Gatebox was shown and explained, however it was not well received by some participants and it was often stressed that they personally do not need the the device. This may be because the target group of gatebox seems to be otakus, which is a term with a negative connotation referring to people highly interested in popculture such as manga, anime or video games including featured virtual characters [45]. Therefore, there is a possibility that participants talked about chatbots and avatars in association with gatebox and otaku culture and perceived them as more negative, thus stressing that they do not want to talk to a virtual companion or human-like avatars.

Not all requirements could be depicted through the prototype. Especially implications for a chatbots dialog were not

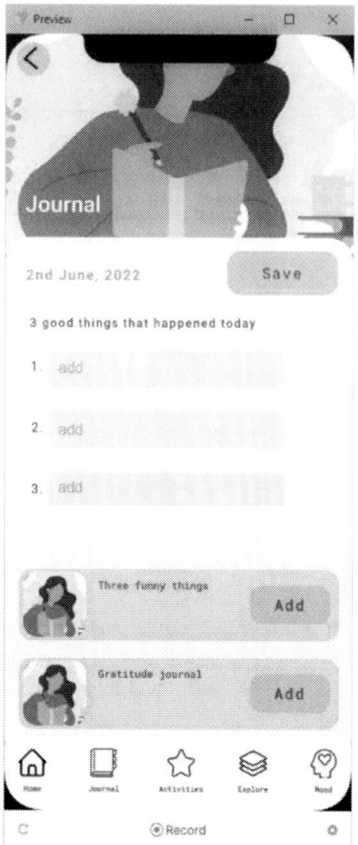

Fig. 1. Journal Exercises

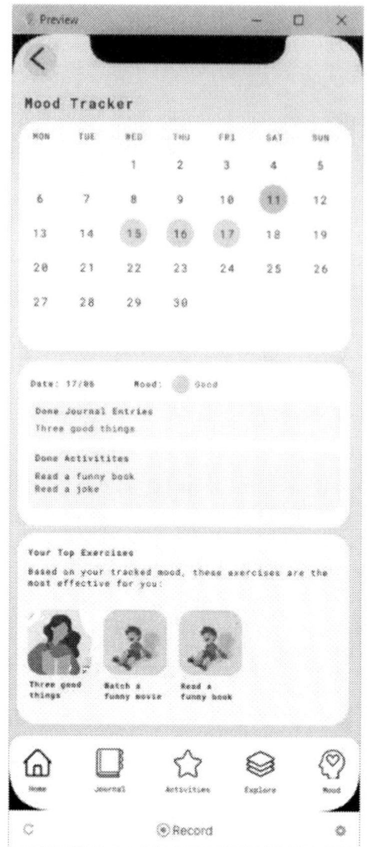

Fig. 2. Mood Tracker

built into a high-fi prototype, such as offering a variety in answers of the chatbot (IA03). Another requirement based on literature was omitted as it contradicted with the interview results: according to previous research, integrating avatars with a similar appearance to users seems to lead to a higher persuasiveness [37]. However, most interviewees stated that they prefer a non-humanlike avatar. This confirms the need to conduct research grounded in case studies [46], [47], and furthermore, at least one additional step of evaluating the prototype should be considered in future work.

There are several limitations to be mentioned. We recognize that Japan is a large country with a high diversity in cultural values and mindsets, which could not be discussed in-depth in this paper. Due to the language barrier, there is the possibility that the literature analysis is missing relevant papers in Japanese language. Furthermore, the interviews were conducted on a smaller scale, and thus interviews with more participants would be useful to obtain more enrich the results in a follow-up step, and including also an evaluation.

VII. Conclusion

In this paper, a prototype for a smartphone application was implemented with the aim to promote mental wellbeing for people with Japanese background living in Europe. As stigma is still prevalent regarding mental health issues in Japanese society, people tend to avoid help-seeking and deal with problems on their own. Therefore, this prototype should provide a tool for self-help and to improve wellbeing through the use of positive psychology. Although a great amount of research regarding positive psychology has been done, it has been mainly tailored to individualistic cultures. This work also aimed at exploring the suitability of positive computing for people for collectivistic cultural backgrounds like Japan, considering its suitability while designing the prototype. Interviews were conducted to specify design implications in detail. Moreover, the usage of chatbots and avatars has been explored and integrated into the application, representing a virtual companion that users can talk with and seek encouragement from. At the same time, the application offers journal exercises and activities to increase positive mood and in addition educational content about mental health and methods to deal with negative emotions. An evaluation of the prototype was not conducted and is something to consider in future work. Also, the integration of wearables or machine learning for automatic mood identification may be an interesting addition. Certainly, more research is needed to diversify the knowledge on positive psychology interventions when it comes to target users with more diverse cultural backgrounds.

REFERENCES

[1] S. Ando, S. Yamaguchi, Y. Aoki, and G. Thornicroft, "Review of mental-health-related stigma in Japan," *Psychiatry and Clinical Neurosciences*, vol. 67, no. 7, pp. 471–482, 2013.

[2] Y. Kotera, P. Gilbert, K. Asano, I. Ishimura, and D. Sheffield, "Self-criticism and self-reassurance as mediators between mental health attitudes and symptoms: Attitudes toward mental health problems in Japanese workers," *Asian Journal of Social Psychology*, vol. 22, no. 2, pp. 183–192, 2019.

[3] K. Yoshioka, N. J. Reavley, L. M. Hart, and A. F. Jorm, "Recognition of mental disorders and beliefs about treatment: results from a mental health literacy survey of Japanese high school students," *International Journal of Culture and Mental Health*, vol. 8, pp. 207–222, Apr. 2015.

[4] R. A. Calvo and D. Peters, *Positive Computing: Technology for Wellbeing and Human Potential*. MIT Press, Nov. 2014. Google-Books-ID: uI6ZBQAAQBAJ.

[5] V. Braun and V. Clarke, "Reflecting on reflexive thematic analysis," *Qualitative Research in Sport, Exercise and Health*, vol. 11, pp. 589–597, Aug. 2019.

[6] E. B. R. Desapriya and I. Nobutada, "Stigma of mental illness in Japan," *The Lancet*, vol. 359, p. 1866, May 2002.

[7] M. Kasahara-Kiritani, T. Matoba, S. Kikuzawa, J. Sakano, K. Sugiyama, C. Yamaki, M. Mochizuki, and Y. Yamazaki, "Public perceptions toward mental illness in Japan," *Asian Journal of Psychiatry*, vol. 35, pp. 55–60, June 2018.

[8] K. Yoshioka, N. J. Reavley, A. J. MacKinnon, and A. F. Jorm, "Stigmatising attitudes towards people with mental disorders: Results from a survey of Japanese high school students," *Psychiatry Research*, vol. 215, pp. 229–236, Jan. 2014.

[9] Y. Kido, N. Kawakami, Y. Miyamoto, R. Chiba, and M. Tsuchiya, "Social Capital and Stigma Toward People with Mental Illness in Tokyo, Japan," *Community Mental Health Journal*, vol. 49, pp. 243–247, Apr. 2013.

[10] A. Masuda, S. C. Hayes, M. P. Twohig, J. Lillis, L. B. Fletcher, and A. T. Gloster, "Comparing Japanese International College Students' and U.S. College Students' Mental-Health-Related Stigmatizing Attitudes," *Journal of Multicultural Counseling and Development*, vol. 37, no. 3, pp. 178–189, 2009.

[11] D. Houri, E. W. Nam, E. H. Choe, L. Z. Min, and K. Matsumoto, "The mental health of adolescent school children: a comparison among Japan, Korea, and China," *Global Health Promotion*, vol. 19, pp. 32–41, Sept. 2012.

[12] K. Ishii, T. Mojaverian, K. Masuno, and H. S. Kim, "Cultural Differences in Motivation for Seeking Social Support and the Emotional Consequences of Receiving Support: The Role of Influence and Adjustment Goals," *Journal of Cross-Cultural Psychology*, vol. 48, pp. 1442–1456, Oct. 2017. Publisher: SAGE Publications Inc.

[13] A. Gaggioli, G. Riva, D. Peters, and R. A. Calvo, "Chapter 18 - Positive Technology, Computing, and Design: Shaping a Future in Which Technology Promotes Psychological Well-Being," in *Emotions and Affect in Human Factors and Human-Computer Interaction* (M. Jeon, ed.), pp. 477–502, San Diego: Academic Press, Jan. 2017.

[14] A. Gaggioli, D. Villani, S. Serino, R. Banos, and C. Botella, "Editorial: Positive Technology: Designing E-experiences for Positive Change," *Frontiers in Psychology*, vol. 10, 2019. Publisher: Frontiers.

[15] M. Seligman, T. Steen, N. Park, and C. Peterson, "Positive Psychology Progress: Empirical Validation of Interventions.," *The American psychologist*, vol. 60, pp. 410–21, July 2005.

[16] S. Wellenzohn, R. T. Proyer, and W. Ruch, "Who Benefits From Humor-Based Positive Psychology Interventions? The Moderating Effects of Personality Traits and Sense of Humor," *Frontiers in Psychology*, vol. 9, 2018.

[17] H. Kim, K. Doiron, M. Warren, and S. Donaldson, "The international landscape of positive psychology research: A systematic review," *International Journal of Wellbeing*, vol. 8, July 2018. Number: 1.

[18] Ando, Hideyuki and Watanabe, Junji and ドミニク, チェン and 一真, 青山 and 杏介, 坂倉, "Wellbeing を促進する情報技術の検討," "横幹連合コンファレンス予稿集", vol. 2017, pp. A–1–3, 2017.

[19] N. Park, C. Peterson, and M. E. P. Seligman, "Strengths of Character and Well-Being," *Journal of Social and Clinical Psychology*, vol. 23, pp. 603–619, Oct. 2004.

[20] A. M. Wood, J. J. Froh, and A. W. A. Geraghty, "Gratitude and well-being: a review and theoretical integration," *Clinical Psychology Review*, vol. 30, pp. 890–905, Nov. 2010.

[21] D. E. Davis, E. Choe, J. Meyers, N. Wade, K. Varjas, A. Z. Gifford, A. Quinn, J. Hook, D. V. V. Tongeren, B. J. Griffin, and E. Worthington, "Thankful for the little things: A meta-analysis of gratitude interventions.," *Journal of counseling psychology*, 2016.

[22] C. Gherghel and T. Hashimoto, "The meaning of kindness and gratitude in Japan: A mixed-methods study," *International Journal of Wellbeing*, vol. 10, Sept. 2020. Number: 4.

[23] K. Layous, H. Lee, I. Choi, and S. Lyubomirsky, "Culture Matters When Designing a Successful Happiness-Increasing Activity: A Comparison of the United States and South Korea," *Journal of Cross-Cultural Psychology*, vol. 44, pp. 1294–1303, Nov. 2013. Publisher: SAGE Publications Inc.

[24] J. K. Boehm, S. Lyubomirsky, and K. M. Sheldon, "A longitudinal experimental study comparing the effectiveness of happiness-enhancing strategies in Anglo Americans and Asian Americans," *Cognition and Emotion*, vol. 25, pp. 1263–1272, Nov. 2011.

[25] K. Otake, S. Shimai, J. Tanaka-Matsumi, K. Otsui, and B. L. Fredrickson, "Happy People Become Happier through Kindness: A Counting Kindnesses Intervention," *Journal of Happiness Studies*, vol. 7, pp. 361–375, Sept. 2006.

[26] F. Gander, R. T. Proyer, W. Ruch, and T. Wyss, "Strength-Based Positive Interventions: Further Evidence for Their Potential in Enhancing Well-Being and Alleviating Depression," *Journal of Happiness Studies*, vol. 14, pp. 1241–1259, Aug. 2013.

[27] S. Wellenzohn, R. T. Proyer, and W. Ruch, "Humor-based online positive psychology interventions: A randomized placebo-controlled long-term trial," *The Journal of Positive Psychology*, vol. 11, pp. 584–594, Nov. 2016.

[28] W. Ruch and P. E. McGhee, "Humor Intervention Programs," in *The Wiley Blackwell Handbook of Positive Psychological Interventions*, pp. 179–193, John Wiley & Sons, Ltd, 2014.

[29] C. Peterson, W. Ruch, U. Beermann, N. Park, and M. Seligman, "Strengths of character, orientations to happiness, and life satisfaction," 2007.

[30] R. Tsukawaki, N. Kojima, T. Imura, Y. Furukawa, and K. Ito, "Relationship between types of humour and stress response and well-being among children in Japan," *Asian Journal of Social Psychology*, vol. 22, no. 3, pp. 281–289, 2019.

[31] H. Tsujita and J. Rekimoto, "Smiling makes us happier: enhancing positive mood and communication with smile-encouraging digital appliances," in *Proceedings of the 13th international conference on Ubiquitous computing - UbiComp '11*, (Beijing, China), p. 1, ACM Press, 2011.

[32] M. V. Birk and R. L. Mandryk, "Improving the Efficacy of Cognitive Training for Digital Mental Health Interventions Through Avatar Customization: Crowdsourced Quasi-Experimental Study," *Journal of Medical Internet Research*, vol. 21, p. e10133, Jan. 2019. Company: Journal of Medical Internet Research Distributor: Journal of Medical Internet Research Institution: Journal of Medical Internet Research Label: Journal of Medical Internet Research Publisher: JMIR Publications Inc., Toronto, Canada.

[33] "Woebot Health," 2022.

[34] A. Darcy, J. Daniels, D. Salinger, P. Wicks, and A. Robinson, "Evidence of Human-Level Bonds Established With a Digital Conversational Agent: Cross-sectional, Retrospective Observational Study," *JMIR Formative Research*, vol. 5, p. e27868, May 2021. Company: JMIR Formative Research Distributor: JMIR Formative Research Institution: JMIR Formative Research Label: JMIR Formative Research Publisher: JMIR Publications Inc., Toronto, Canada.

[35] "Woebot: your self-care expert in CBT & mindfulness – google play reviews," 2022.

[36] "Woebot: Your self-care expert appstore reviews," 2022.

[37] M. Franco, C. Monfort, A. Piñas-Mesa, and E. Rincon, "Could Avatar Therapy Enhance Mental Health in Chronic Patients? A Systematic Review," *Electronics*, vol. 10, p. 2212, Jan. 2021. Number: 18 Publisher: Multidisciplinary Digital Publishing Institute.

[38] I. C. Rehm, E. Foenander, K. Wallace, J.-A. M. Abbott, M. Kyrios, and N. Thomas, "What Role Can Avatars Play in e-Mental Health Interventions? Exploring New Models of Client–Therapist Interaction," *Frontiers in Psychiatry*, vol. 7, 2016.

[39] "Gatebox inc.," 2022.

[40] D. White and P. W. Galbraith, "Japan's Emerging Emotional Tech," *Anthropology News*, vol. 60, Jan. 2019.

[41] I. Schwaninger, C. Frauenberger, and G. Fitzpatrick, "Unpacking Forms of Relatedness around Older People and Telecare," in *DIS' 20 Companion: Companion Publication of the 2020 ACM Designing Interactive Systems Conference*, pp. 163–169, New York, NY, USA: Association for Computing Machinery, July 2020.

[42] P. Sripian, M. N. A. M. Anuardi, J. Yu, and M. Sugaya, "The Implementation and Evaluation of Individual Preference in Robot Facial Expression Based on Emotion Estimation Using Biological Signals," *Sensors*, vol. 21, p. 6322, Sept. 2021.

[43] I. Schwaninger, F. Güldenpfennig, A. Weiss, and G. Fitzpatrick, "What Do You Mean by Trust? Establishing Shared Meaning in Interdisciplinary Design for Assistive Technology," *Int. J. Social Rob.*, vol. 13, pp. 1879–1897, Dec. 2021.

[44] S. Lee, H. Kim, M. J. Park, and H. J. Jeon, "Current Advances in Wearable Devices and Their Sensors in Patients With Depression," *Front. Psychiatry*, vol. 12, June 2021.

[45] E. Civil, "Otaku: Japanese 'obsessive' subculture explained," n.a.

[46] H. R. Lee and S. Šabanović, "Weiser's dream in the Korean home: collaborative study of domestic roles, relationships, and ideal technologies," in *UbiComp '13: Proceedings of the 2013 ACM international joint conference on Pervasive and ubiquitous computing*, pp. 637–646, New York, NY, USA: Association for Computing Machinery, Sept. 2013.

[47] B. M. Nur, "A case study of socio-cultural and technical factors in automobile design: Discourses between designers and potential users on a new electric vehicle in Africa," *Technology in Society*, vol. 63, p. 101398, Nov. 2020.

A Secure Blockchain for Electronic Health Records

Jihad Qaddour
School of Information Technology
Illinois State University
Old Union Building, Box 5150, Normal, IL 61790, USA
jqaddou@ilstu.edu

Kanz Ul Eman)
School of Information Technology
Illinois State University
Old Union Building, Box 5150, Normal, IL 61790, USA
keman@ilstu.edu

Abstract—**Blockchain technology has attracted considerable attention and has grown constantly since its introduction in 2008. It has emerged as a valuable tool in several industries, including healthcare, particularly regarding protecting and securing electronic health records. They contain sensitive patient data that is often vulnerable to cyberattacks. Blockchain's decentralized and immutable nature can help to protect EHRs from unauthorized access, modification, or deletion. This paper proposes a blockchain-based architecture for EHRs that incorporates Ethereum-based smart contracts, decentralized off-chain storage with the Interplanetary File System, and strong symmetric encryption. This architecture provides a robust solution that guarantees the security and scalability of EHRs. The paper also provides a thorough analysis of the framework's security merits and improves our knowledge and ability to use secure electronic health record systems.**

Keywords- Blockchain; health data; electronic health records; security; decentralize; confidentiality.

I. INTRODUCTION

The healthcare industry generates and maintains a vast amount of data daily, including highly sensitive information such as medical records, diagnoses, vital signs, and drug regimens [1]. Electronic health records (EHRs) have become widely adopted in the healthcare industry due to significant technological advances. In fact, by 2017, 96% of non-federal acute hospitals in the United States had adopted EHR systems [2]. This adoption demonstrates the industry's understanding of the benefits of EHRs in improving data management and accessibility, optimizing workflows, and improving patient care.

EHRs and well-established health information exchange (HIE) systems work together to provide a number of benefits, including lower healthcare costs and higher care quality [3]. However, using digital technologies to transmit highly private and sensitive information raises concerns, particularly about privacy and security. When information is shared between different healthcare organizations, there is an increased risk of unauthorized access, making it vulnerable to potential hackers who could launch targeted attacks [4].

To address these concerns, it is essential to create strong security controls and privacy protections. Solutions such as encryption, access controls, secure authentication protocols, and data anonymization techniques can be used to preserve the confidentiality and integrity of shared data [4]. Secure sharing of health information can also be ensured through ongoing monitoring, frequent security assessments, and employee training. These measures can also help to mitigate risks.

This research proposes a novel decentralized approach to enhancing the security of electronic health records (EHRs). The proposed system provides a scalable and decentralized alternative for storing and distributing EHR data using the Ethereum blockchain and the Interplanetary File System (IPFS) [4]. This approach leverages the built-in security features of blockchain technology and the durability of IPFS to guarantee the integrity and privacy of private medical records. The project's focus on decentralization is coherent with the growing demand in the healthcare industry for secure and effective EHR storage and interchange.

The rest of this paper is organized as follows. Section II provides background and related work. Section III discusses the proposed model. Section IV concludes the paper with the future direction.

II. BACKGROUND AND RELATED WORK

The healthcare industry is facing a critical challenge in ensuring the security of information flow. Data breaches in the healthcare industry have affected millions of people in recent years, highlighting the need for strong security measures to safeguard sensitive healthcare data [5][6]. Blockchain technology is a promising solution for information exchange in the healthcare industry. Blockchain uses a distributed ledger to ensure that every participating node keeps an exact copy of the ledger, improving data integrity and transparency [6]. Blockchain is also decentralized and irreversible, which makes it ideal for secure data sharing [6]. While blockchain technology presents exciting possibilities for electronic health records (EHR) in the healthcare sphere, there are significant challenges to overcome. Scalability is a major challenge because the bulk of EHR data can be enormous, resulting in slower and longer transactions when stored on the blockchain [6]. Another issue is transparency, as all transactions on a blockchain are public, posing privacy issues for sensitive healthcare data [6]. Balancing the need for privacy while guaranteeing effective blockchain tracking and recordkeeping becomes a critical challenge [6].

There is a growing body of research on how to overcome the challenges of using blockchain technology in healthcare. Matos et al. [7] presented a system design that uses cloud services and granular access control to successfully administer EHR. The goal was to create a safe and scalable solution that allows patients and clinicians to access EHR from anywhere in the world. Intercloud storage was used, which entails joining separate clouds to form a bigger network, allowing for end-to-end anonymity and smooth data migration between providers. In their access control method, Matos et al. emphasized the need for authentication and permission checks. However, despite efforts to prioritize patient privacy, the system may still be open to exploitations that could allow unauthorized access to critical data [7].

A. The Hyperledger Fabric blockchain

The Hyperledger Fabric blockchain, a private blockchain system, is used by the proposed framework, Action-EHR, presented by Dubovitskaya et al. [8] to improve authentication and authorization procedures. Hyperledger Fabric offers tighter control over node involvement and transaction visibility than open public blockchains. This system uses smart contracts to manage access control and

preserve state variables pertaining to patient health records, much like the Ethereum network. Action-EHR intends to provide secure and auditable access to patient data by utilizing the smart contract logic. Fine-grained control over access permissions is made possible using Hyperledger Fabric and smart contracts, improving data privacy and security [8]. This strategy supports ongoing research into the effective and safe management of electronic health records using blockchain technology [8]. The misconception that blockchain technology was first developed for cryptocurrencies is a common one. However, the idea of blockchain was first proposed in 1991 [10]. The original concept behind blockchain was to create a system for digital document timestamping to prevent manipulation or backdating. This suggests that blockchain has applications beyond cryptocurrencies and can be used in various industries that require secure record-keeping and transaction tracking [10].

Distributed ledgers are used by blockchain technology to record and keep all transactions made on the network. An immutable hash signature is present in each block to which a transaction is added. Data manipulation within the blockchain is typically impossible due to the decentralized structure of the blockchain network and this hash, which assures that any unauthorized modifications to the blocks would be instantly identified and rejected [6].

B. Ethereum with Smart Contracts

Blockchain technology has made significant improvements since it was first used in Bitcoin. In 2015, Ethereum joined Bitcoin as a prominent cryptocurrency, building on the research report written by Vitalik Buterin two years prior [11]. A new blockchain was introduced by Ethereum that was comparable to Bitcoin's but distinguished itself by including smart contracts. Smart contracts allow logical code to be executed directly on the ledger, expanding the capabilities of the blockchain beyond basic transactions [11]. This innovation by Ethereum helped blockchain technology become more widely used and developed, making it a promising solution for information exchange in the healthcare industry.

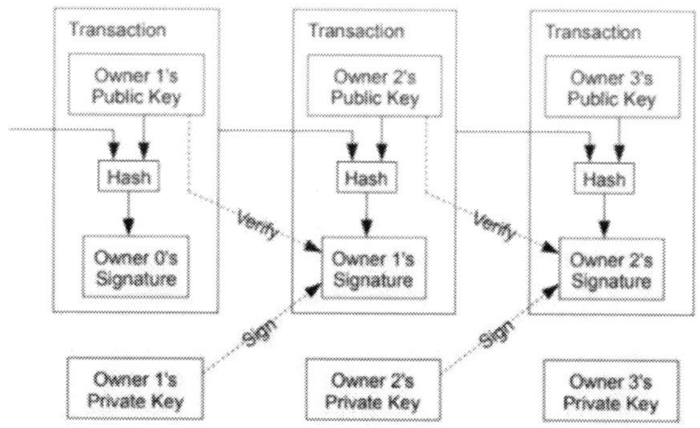

Figure 1. Process of hash signing [6]

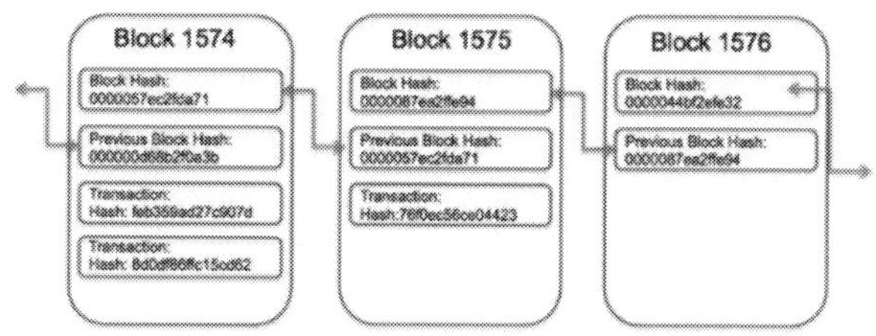

Figure 2: Connections of the blocks [9]

C. IPFS: A Decentralized File Access and Storage Protocol

Interplanetary File System (IPFS)) is a peer-to-peer hypermedia protocol that provides a decentralized file access and storage system. IPFS differs from previous peer-to-peer protocols such as Bit Torrent in that it uses a content-addressable addressing scheme. This means that data is divided into manageable portions, hashed, and assigned a CID (Content Identifier) value. This special addressing scheme ensures data integrity and reduces duplication, which enables efficient file retrieval on the IPFS network [12].

D. MedRec: A Blockchain-Based Electronic Health Records Platform

MedRec is a cutting-edge electronic health records (EHR) platform that is built on the Ethereum network. MedRec uses smart contracts written in Python to manage access to and permissions for EHR data. MedRec also features an innovative incentive system that rewards healthcare practitioners for contributing anonymized medical data. This incentive system helps to build trust and facilitate access to valuable healthcare information [13].

E. Ancile and BHEEM

In addition to MedRec, there are several other blockchain-based EHR platforms that have been proposed. These platforms share many similarities with MedRec, in that they all use blockchain technology to provide a secure and decentralized way to store and manage EHR data, including Ancile [14] and BHEEM 15]. Both frameworks use the Ethereum network/blockchain for access management and permissions while keeping health records off-chain in a local database. While Ancile uses two encryption techniques for record storage and distribution, BHEEM omits a specific description of the encryption method used. Asymmetric encryption and proxy re-encryption are used in the distribution encryption in the Ancile framework. This enables the restoration of fully encrypted messages using a user's private key, even if the encryption was carried out using a different user's public key [16].

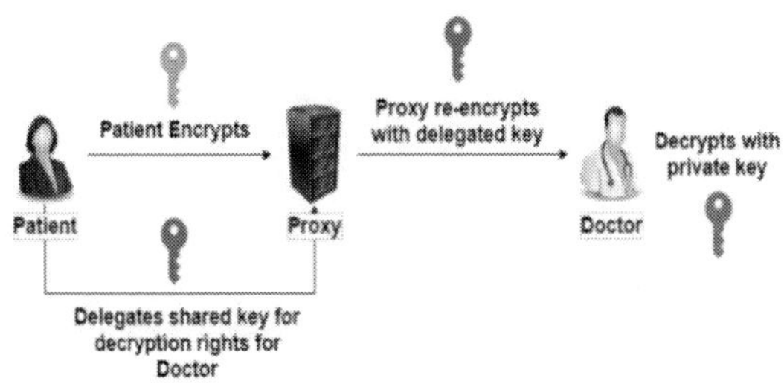

Figure 3. An example of proxy re-encryption [16].

F. Patient-centric framework for personal health records (PHR)

Madine et al. [17] developed a patient-centric framework for personal health records (PHR) using blockchain technology, specifically the Ethereum network. The framework uses smart contracts to create an access control system. IPFS and proxy re-encryption are used as complementary techniques to overcome scalability issues. Madine et al.'s [17] research included a thorough comparison between their blockchain-based PHR architecture and current cloud-based PHR solutions. The comparison focused on a number of factors, including provenance, immutability, trustworthiness, patient-centered approach, decentralized storage, decentralized execution, and privacy. A full comparison of the results is presented in Table 1 of their research.

TABLE 1. COMPARISON OF BLOCKCHAIN-BASED PHR ARCHITECTURE WITH CLOUD-BASED PHR SOLUTIONS.

Factor	Blockchain-Based PHR	Cloud-Based PHR
Provenance	All data modifications are tracked and recorded on the blockchain, providing a complete audit trail.	Data modifications are typically not tracked or recorded in the cloud, making it difficult to audit changes to data.
Immutability~	Once data is added to the blockchain, it cannot be modified or deleted.	Data in the cloud can be modified or deleted at any time by the cloud provider or by authorized users.
Trustworthiness	The blockchain is a decentralized network, so there is no single point of failure or control.	Cloud-based PHR solutions are typically centralized, which means that there is a single point of failure and control.
Patient-Centered Approach	Patients have complete control over their data and who has access to it.	Patients typically do not have complete control over their data in cloud-based PHR solutions.
Decentralized Storage	Data is stored on multiple nodes on the blockchain, making it more secure and resistant to data breaches.	Data is typically stored on a single server in the cloud, which makes it more vulnerable to data breaches.
Decentralized Execution	Smart contracts are executed on the blockchain, which ensures that they are tamper-proof and cannot be censored.	Smart contracts are typically executed on a centralized server in the cloud, which makes them vulnerable to tampering and censorship.
Privacy	Data can be encrypted on the blockchain, which can help to protect patient privacy.	Data in the cloud is typically not encrypted, which makes it more vulnerable to unauthorized access.

III. PROPOSED MODEL

The proposed system uses the Ethereum blockchain to store and manages electronic health records (EHRs). Ethereum is an open-source platform for smart contracts and decentralized applications. Smart contracts are self-executing contracts that are stored on the blockchain and cannot be tampered with. Figure 4 shows the architecture of the proposed model. The proposed architecture is secure in terms of confidentiality, security, and integrity. Patients have full control over their EHRs and who has access to them. The secret keys used to encrypt EHRs are randomly generated and not user-dependent, which provides a defense against brute-force attacks. The blockchain provides integrity by ensuring that EHRs cannot be tampered with.

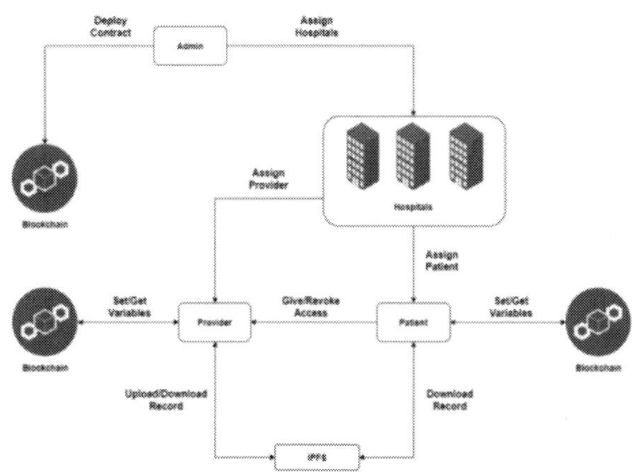

Figure 4. Architecture of the proposed model

The proposed architecture consists of four types of nodes: administrator, hospital, patient, and **provider,** as illustrated in Figure 4. The system starts with the administrator node, which is maintained and owned by the system developer. The **administrator node** is responsible for managing the system and connecting hospital nodes to the blockchain. **Hospital nodes** are responsible for issuing Ethereum addresses to patients and providers. **Patient nodes** store patient identifiers, such as gender and Ethereum address, as well as a mapping of which providers have access to the patient's EHR. Therefore, the patient is still in charge of keeping this mapping up to date, authorizing or denying access to providers as needed, and preserving control over their private medical data. **Provider nodes** store provider identifiers, such as Ethereum address and specialty. This node can create and edit health records, as well as access an existing patient's health record. However, it is the patient's obligation to add the provider to their access list, providing them permission to read and interact with the patient's health records.

The following are the steps involved in the operation of the proposed architecture:

1. The patient authorizes the creation of an EHR by a particular provider.
2. The authorized provider creates the patient's EHR.
3. The EHR is uploaded to the IPFS-distributed file storage system.
4. The EHR is encrypted using the Advanced Encryption Standard (AES) symmetric encryption algorithm.
5. The hash of the encrypted EHR is stored on the blockchain.
6. The patient or provider enters the hash to access the EHR.
7. Smart contracts are included in the framework to create the access control system. Only authorized parties can see and edit health records thanks to the execution of access control restrictions which are made possible by the incorporation of smart contracts.
8. Smart contracts can be created using Ethereum-specific programming languages such as Solidity programming or Python.
9. The encrypted EHR is downloaded using the hash.
10. The EHR is decrypted using the shared encryption key generated during the encryption process.
11. The encryption key is kept private and is only securely shared with the patient and authorized providers who need access to the EHR.

The proposed architecture provides a secure and scalable mechanism for managing EHRs. It is patient-centric, giving patients full control over their data. It is also tamper-proof, ensuring that EHRs cannot be modified without the patient's consent.

The proposed architecture uses blockchain, smart contracts, and decentralized storage to improve the security, integrity, scalability, and access control of EHRs.

The proposed architecture has several **advantages** over traditional EHR systems, including:

- Security: Blockchain provides a high level of security for EHRs by making them tamper-proof and immutable.
- Integrity: Decentralized storage ensures that EHRs are not lost or corrupted.
- Scalability: The blockchain can be scaled to support many users and transactions.
- Access control: Smart contracts can be used to enforce access control policies for EHRs.

IV. CONCLUSION

In this paper, a framework for the blockchain-based management of electronic health records (EHR) is presented. The proposed architecture uses the Ethereum blockchain, smart contracts, and decentralized storage systems like IPFS to address the issues of privacy, security, scalability, and access control in the healthcare industry.

The architecture improves the security and integrity of EHR by utilizing blockchain's distributed ledger, immutable transactions, and cryptographic techniques. This makes it difficult for unauthorized parties to access or modify EHR data. Additionally, the architecture enables fast and transparent data sharing by allowing authorized parties to view and edit EHR data. This is made possible using smart contracts to enforce access control restrictions.

Decentralized storage solutions also increase data availability and lower the risk of data loss or tampering. This is because EHR data is stored on multiple nodes in the decentralized network, making it more difficult to lose or alter.

The patient-centric approach of the framework gives individuals control over their own health information. This includes the ability to store, view, and share their EHR data with authorized parties.

The architecture works as follows:

1. Patients create their own EHRs and store them on the blockchain.
2. Providers can access patients' EHRs if the patient has granted them permission.
3. Access control is enforced using smart contracts.
4. EHRs are stored in a decentralized storage system, making them more secure and available.

The proposed architecture appears to be promising in terms of resolving the problems of traditional EHR systems. However, more research and evaluation are needed to demonstrate its effectiveness, scalability, and real-world applicability.

REFERENCES

[1] The National Coordinator for Health Information Technology, "What information does an electronic health record (ehr) contain," [Online] Available: https://www.healthit.gov/faq/what-electronic-health-record-ehr, 2019.

[2] "Non-federal acute care hospital electronic health record adoption," Health IT Quick-Stat 47, 09 2017, [Online] Available: https://dashboard.healthit.gov/ quickstats/pages/FIG-Hospital-EHR-Adoption.php, 2017.

[3] N. Menachemi, S. Rahurkar, C. A. Harle, and J. R. Vest, "The benefits of health information exchange: an updated systematic review," Journal of the American Medical Informatics Association, vol. 25, no.

9, pp. 1259–1265, 04 2018. [Online]. Available: https://doi.org/10.1093/jamia/ocy035, April 2018.

[4] J. Goodman, L. Gorman, and D. Herrick, "Health information technology: Benefits and problems," 2010. [Online] Available: https://www.ncpathinktank.org/pdfs/st327.pdf, 2010.

[5] P. R. Clearinghouse, "Data breaches." [Online] Available: https://privacyrights.org/ data-breaches.

[6] S. Nakamoto, "Bitcoin: A peer-to-peer electronic cash system," 2009. [Online] Available: http://www.bitcoin.org/bitcoin.pdf.

[7] D. R. Matos, M. L. Pardal, P. Adão, A. R. Silva, and M. Correia, "Securing electronic health records in the cloud," in Proceedings of the 1st Workshop on Privacy by Design in Distributed Systems, ser. W-P2DS'18. New York, NY, USA: Association for Computing Machinery, 2018. [Online] Available: https://doiorg.proxy.lnu.se/10.1145/3195258.3195259, 2018.

[8] A. Dubovitskaya, F. Baig, Z. Xu, R. Shukla, P. S. Zambani, A. Swaminathan, M. M. Jahangir, K. Chowdhry, R. Lachhani, N. Idnani, M. Schumacher, K. Aberer, S. D. Stoller, S. Ryu, and F. Wang, "Action-ehr: Patient-centric blockchain-based electronic health record data management for cancer care," J Med Internet Res, vol. 22, no. 8, p. e13598, Aug 2020. [Online] Available: http://www.jmir.org/2020/8/e13598/, 2020.

[9] M. Gupta, Blockchain For Dummies, 3rd IBM Limited Edition. John Wiley & Sons, Inc, doi:10.1126/science.1065467, Dec. 2020.

[10] S. Haber and W. S. Stornetta, "How to time-stamp a digital document," J. Cryptol., vol. 3, no. 2, p. 99–111, Jan. 1991. [Online] Available: https://doi-org.proxy.lnu.se/10.1007/BF00196791, 1991.

[11] V. Buterin, "Ethereum: A next-generation smart contract and decentralized application platform," 2013. [Online] Available: https://github.com/ethereum/wiki/ wiki/White-Paper, 2013.

[12] J. Benet, "IPFS - content addressed, versioned, P2P file system," CoRR, vol. abs/1407.3561, 2014. [Online] Available: http://arxiv.org/abs/1407.3561, 2014.

[13] A. Ekblaw, A. Azaria, J. D. Halamka, and A. Lippman, "A case study for blockchain in healthcare:"medrec" prototype for electronic health records and medical research data," in Proceedings of IEEE Open & big data conference, vol. 13, 2016, p. 13.

[14] G. G. Dagher, J. Mohler, M. Milojkovic, and P. B. Marella, "Ancile: Privacy-preserving framework for access control and interoperability of electronic health records using blockchain technology," Sustainable Cities and Society, vol. 39, pp. 283–297, 2018. [Online] Available: https://www.sciencedirect.com/science/article/ pii/S2210670717310685, 2018.

[15] J. Vora, A. Nayyar, S. Tanwar, S. Tyagi, N. Kumar, M. S. Obaidat, and J. J. P. C. Rodrigues, "Bheem: A blockchain-based framework for securing electronic health records," in 2018 IEEE Globecom Workshops (GC Wkshps), pp. 1–6, 2018.

[16] T. Matsuo, "Proxy re-encryption systems for identity-based encryption," in PairingBased Cryptography – Pairing 2007, T. Takagi, T. Okamoto, E. Okamoto, and T. Okamoto, Eds. Berlin, Heidelberg: Springer Berlin Heidelberg, pp. 247– 267, 2007.

[17] M. M. Madine, A. A. Battah, I. Yaqoob, K. Salah, R. Jayaraman, Y. Al-Hammadi, S. Pesic, and S. Ellahham, "Blockchain for giving patients control over their medical records," IEEE Access, vol. 8, pp. 193 102–193 115, 2020.

APPENDIX A

Simple outline for a secure Algorithm for Blockchain-Based Architecture for HER in Python

```
def create_ehr(patient, provider):
  # Create a new EHR for the patient.
  # The provider must be authorized to create the EHR.
  ehr = {
    "patient": patient,
    "provider": provider,
  }
  # Upload the EHR to IPFS.
  ehr_hash = ipfs.upload_file(ehr)
  # Encrypt the EHR using AES.
  encryption_key = generate_encryption_key()
  encrypted_ehr = encrypt_ehr(ehr, encryption_key)
  # Store the hash of the encrypted EHR on the blockchain.
  blockchain.store_hash(ehr_hash)
  # Create a smart contract to manage access to the EHR.
  smart_contract = create_smart_contract()
  # Add the patient and authorized providers to the smart contract.
  smart_contract.add_user(patient)
  for provider in ehr["authorized_providers"]:
    smart_contract.add_user(provider)
  # Share the encryption key with the patient and authorized
providers.
  patient.set_encryption_key(encryption_key)
  for provider in ehr["authorized_providers"]:
    provider.set_encryption_key(encryption_key)
def access_ehr(patient, hash):
  # Check if the patient is authorized to access the EHR.
  if not smart_contract.is_user_authorized(patient):
    raise UnauthorizedAccessError()
  # Download the encrypted EHR from IPFS.
  encrypted_ehr = ipfs.download_file(hash)
  # Decrypt the EHR using the shared encryption key.
  ehr = decrypt_ehr(encrypted_ehr, patient.get_encryption_key())
  # Return the EHR.
  return.
```

HealthSonar: A System for Unobtrusive Monitoring of Elders and Patients with Movement Disorders

Adamantios Ntanis*, Spyridon Kontaxis*, George Rigas*, Anastasia Pentari†, Kostas Tsiouris‡, Efstathios Kontogiannis*, Eleftherios Kostoulas*, Ilias Tsimperis*, Theodoros Vlioras*, Aristotelis Bousis*, Styliani Zelilidou‡, Kalypso Tasiou‡, Manolis Tsiknakis†§ and Dimitrios Fotiadis¶

*PD Neurotechnology Ltd., Ioannina, Greece, GR 455 00
†Institute of Computer Science, Foundation for Research and Technology—Hellas, Heraklion, Greece, GR 700 13
‡Biomedical Research Institute, Foundation for Research and Technology—Hellas, Ioannina, Greece, GR 455 00
§Department of Electrical and Computer Engineering, Hellenic Mediterranean University, Heraklion, Greece, GR 710 04
¶Unit of Medical Technology and Intelligent Information Systems, University of Ioannina, Ioannina, Greece, GR 451 10

Corresponding author: Adamantios Ntanis, e-mail: a.ntanis@pdneurotechnology.com

***Abstract*—This work presents the HealthSonar, a novel unobtrusive, privacy-focused, health monitoring system for elders and patients with movement disorders, capable of tracking the quality of their sleep, identifying sleep apnea events, evaluating their mobility and detecting falls. The system comprises a impulse radio ultra-wideband (IR-UWB) radar-based device, a web portal, a web dashboard and a mobile application. Due to its nature, HealthSonar is perfectly suited for care homes, and generally for clinical environments, where residents are in need of continuous monitoring, especially during their sleep, as well as during their presence in hazardous rooms, such as bathrooms.**

***Index Terms*—ultra-wideband radar; health monitoring; sleep monitoring; gait analysis; fall detection**

I. INTRODUCTION

The world's older population is rising, along with the number of patients suffering from sleep, neurological and movement disorders [1]–[4]. Impaired mobility, in particular, can result in falls, which in turn can result in life-threatening injuries, especially in older people. A practical solution, offering a unified approach for monitoring sleep, evaluating gait and identifying falls, could be of great interest, specifically for care homes and clinical environments. As of now, to the best of our knowledge, there is no such practical unified solution offering all those features.

Lately, progress in ultra-wideband (UWB) radar technology has resulted in the development of affordable, practical and accurate radar sensors [5] that can be used as a platform for building health monitoring solutions. Specifically, because of their high accuracy, penetrating capabilities and reliability, UWB radars can track micro and macro motions, even through different weather conditions and obstacles (such as walls and furniture), making them ideal for a diverse set of applications. Some of them are, respiratory and heart rate extraction, sleep monitoring, presence detection, fall detection, people counting, gesture recognition, baby monitoring, assisted living of elderly people, as well as mobility monitoring [5], [6]. What is more, UWB radar technology provides an unobtrusive, contactless, low-consumption and privacy-focused approach, perfectly suited for devices meant to be used for long periods of time, inside the most private areas of people's homes, such as bedrooms, bathrooms, etc. Due to their nature, radar-based devices are a suitable choice for applications pertaining to sleep, where a wearable solution would inevitably cause discomfort, while a camera-based one would be privacy-intrusive. Moreover, regarding sleep, radars offer unparalleled opportunities for contactless sleep monitoring as their penetrating abilities result in posture recognition even in low temperatures where people would be covered with blankets [7]. Given radar sensors devices are constantly emitting energy in the form of radiation, a logical concern would be that of safety, regarding their continual use. Thankfully, UWB radars are safe to use, due to their low emitted power levels of non-ionizing radiation that are harmless to human health [5].

In this work, we present a novel health monitoring system, the HealthSonar, based on UWB radar technology. The system itself can be used easily in a plethora of clinical scenarios and even as a general tracker for improving the quality of life of healthy people, but its applicability in clinical environments and care homes is particularly noteworthy. Specifically, in Section II the various hardware and software parts of the HealthSonar system are presented and in Section III its functionality is described. Next, in Section IV the innovative aspects of the system are discussed along with possible future directions for its improvement. Last but not least, in Section V some concluding remarks are presented.

II. THE HEALTHSONAR SYSTEM

The HealthSonar system was created to fill the gap in accurate, unobtrusive, privacy-focused, continuous and particularly contactless, health monitoring. More specifically, the system was created to serve elders (especially those with deteriorated mobility) and patients with movement or sleep disorders in tracking the quality of their sleep, identifying sleep apnea events, evaluating their mobility and detecting fall events.

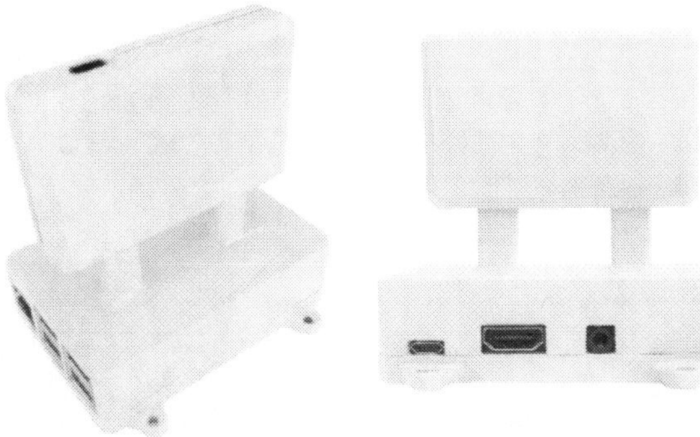

Fig. 1. A prototype of the HealthSonar device.

Moreover, the system also detects human presence, which is a prerequisite and important step for choosing when the sleep monitoring or the fall detection pipelines will initiate.

The development of the system was propelled by the recent advances in IR-UWB radar technology, especially the small sizes of the available sensors and their affordable prices. Those features rendered them commercially viable options to be used as platforms for building health monitoring devices. The HealthSonar system consists of the following "components": (1) a radar-based monitoring device, (2) a web dashboard application, (3) a web portal to the radar device, (4) a mobile application (or app), (5) a cloud data processing service, (6) a cloud data storage service, and (7) an API for communication purposes.

The system was developed as an ecosystem with different users in mind, such as elders, patients, nurses, care workers or attending physicians. It was also designed to provide different administrative and monitoring utilities to each one of the aforementioned groups. The HealthSonar system is meant to be straightforward and easy to use, while still being customizable for advanced, mainly research, use cases. It is important to note that HealthSonar was built as a centralized, extensible system, as a result multiple devices can be connected and administrated at once, in order to cover the needs of a care home, or generally of a clinical environment. On the other hand, the system is modular enough to be usable as a single unit, ideal for home use applications, or even research purposes. Last but not least, HealthSonar is operational with minimal user interaction and after its first setup, it can continually monitor either a bed for sleeping individuals or a bathroom for human activity. Given its lack of a battery, as it will be described below, the system does not even need the users to manually check its power level at regular intervals.

A. The radar-based monitoring device

The cornerstone of the HealthSonar system is the radar-based monitoring device, which was designed, developed and produced, in-house, by PD Neurotechnology Ltd. You can see the prototype of the HealthSonar monitoring device in Fig. 1. The front side of the device can be seen on the left half of Fig. 1, while the back side can be seen on the right half. The case for the prototype was produced using additive manufacturing techniques. The device comprises an Aria Sensing LT102 IR-UWB radar sensor, embedded within the top part, and a Raspberry Pi 4 Model B board encased within the bottom base. The specifications of the Aria Sensing radar module can be found in Table I. On the top, back and side of the device, various connection ports can be seen. The radar connects to the Raspberry Pi board via 1 external USB cable.

TABLE I
ARIA SENSING LT102 RADAR SPECIFICATIONS.

General specifications	Values
Radar's operating frequency	6.5 GHz to 8.5 GHz
Temperature operating range	$-40\,^{\circ}$C to $85\,^{\circ}$C
Radar module's dimensions	36 mm×68 mm
Maximum power consumption	220 mW at 5 V
Integrated antenna aperture	$\pm 60^{\circ}$ by $\pm 60^{\circ}$
Typical detection range	12 m

The radar sensor identifies micro motions (such as the oscillation of the chest) and macro motions (such as the human gait) translating them into data, while the Raspberry Pi board is the main processing unit of the device. The HealthSonar device is not powered by a battery, but through a wall socket, as a result it can work indefinitely without any user interaction for charging purposes. The intended placement of the device is: (1) next to a bed, on top of a nightstand, enabling nighttime sleep monitoring, (2) mounted on a bathroom wall, enabling fall detection, or (3) mounted on top of a tripod, or other suitable furniture, for general mobility evaluation. The system's intended use is depicted in Fig. 3 and includes 2 HealthSonar devices, one placed on top of a nightstand and one on the bathroom wall.

B. The web dashboard application

The web dashboard (see Fig. 2a) is a web application that runs on the cloud and can be accessed through the Internet. Its purpose is to act as the central hub for managing multiple (or more aptly, at least one) connected HealthSonar devices. The dashboard is meant to be used by care home workers or nurses as an administrative and overview tool.

The functionality of the dashboard is: (1) Manually initiating and terminating recording sessions. (2) Manually selecting data for upload to the cloud storage. (3) Accessing telemetry data (logs) for connected devices. (4) Accessing system information (metadata) about each device. (5) Assigning users (e.g., elders, patients) to specific devices. (6) Viewing the connection status of each available device. (7) Managing the recorded stored data of all connected devices. (8) View reports and notifications for the extracted health analytics.

C. The web portal to the radar

The web portal to the radar (see Fig. 2b) is a web application that runs locally on the HealthSonar device itself and can only be accessed through the local network that the device is

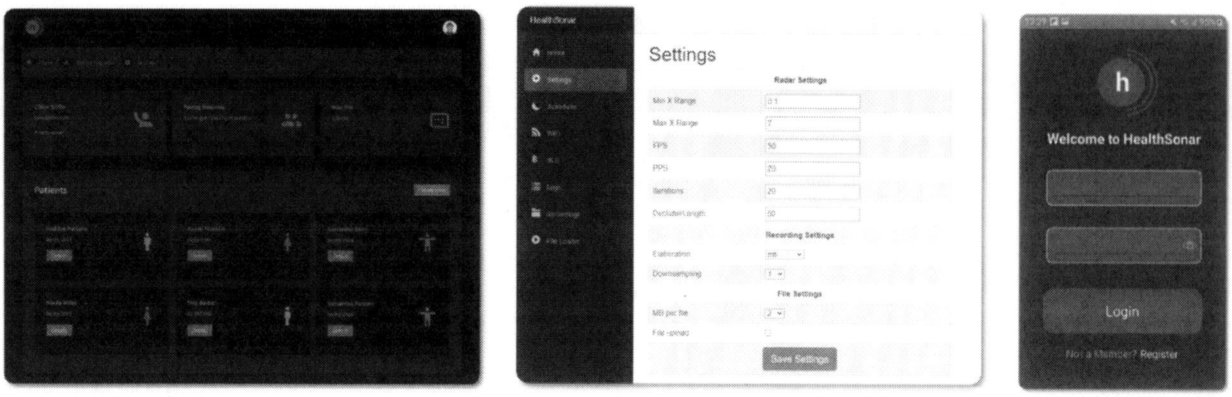

(a) (b) (c)

Fig. 2. (2a) The main screen of the web dashboard. (2b) The settings page of the web portal. (2c) The mobile app's login screen.

connected to, without the need for an Internet connection. Its purpose is to facilitate direct communication with, and access to, one specific HealthSonar device, while also serving as the main configuration tool of its settings.

Generally, the web portal shares a similar purpose with the mobile application, that will be described below (see Section II-D), although the latter is not able to configure the radar. The portal is meant to be used by care workers or nurses, that have been trained in its use, as a tool for setting up the radar device, for rapid implementation of various testing scenarios and data gathering (e.g., sleep or mobility evaluation scenarios), or for easy access to stored radar data.

Specifically, the functionality of the web portal is: (1) Setting up the radar through Wi-Fi. (2) Configuring the settings of a device. (3) Initiating and terminating a recording. (4) Setting up a scheduler for a recording. (5) Accessing previously-stored radar data. As a result, the web portal is intended as a tool for advanced, trained users, that understand the consequences of having direct access to the HealthSonar device's settings and storage, and specifically the effects of changing the internal parameters of the radar.

An example of an ideal use case for the web portal would be that of a clinical trial, or generally any research endeavor, in which the portal would be the main tool of the researchers for configuring the radar based on the needs of specialized scenarios, as well as for accessing the resulted data. It should be noted that, under normal circumstances, the web portal should not be used, as the HealthSonar device is delivered to a user preconfigured. As a result, no tinkering with advanced settings is needed, while the HealthSonar's full functionality can be achieved with the user-friendlier web dashboard and mobile application. This is the main reason that the web portal's user interface is not as polished as the web dashboard's as its target user group.

D. The mobile application (mobile app)

The mobile application (see Fig. 2c) facilitates the initial setup of the system and provides, to the end user, sleep and mobility extracted metrics based on data acquired with the HealthSonar device, enables the user to evaluate their mobility and receives notifications for fall events. The end users of the mobile app are the ones being monitored by the HealthSonar device (usually those are the elders and patients). The functionality of the mobile app, apart from setting up the system during its first use, revolves around providing detailed analyses about sleep quality (through sleep staging and sleep duration), sleep-related events (such as obstructive sleep apnea), mobility metrics indicating deteriorated gait, or a higher fall risk, as well as real-time notifications in the event a fall takes place within the bathroom. Note that the aforementioned presented mobility metrics are calculated using data generated through a TUG test, which the users need to manually initiate through the HealthSonar mobile app itself. As a result, the initiation of a TUG test session is one more feature of the mobile application.

Contrary to feature-rich mobile applications for commercial sleep and fitness trackers, the HealthSonar app is built around user-friendliness and targeted functionality, taking into account, first and foremost, the needs and experience of their users, those being elders and patients. Said groups of people, are not accustomed to technology, or they are not capable of operating an app extensively due to impaired dexterity (for example patients with Parkinson's disease). As a result, in such cases, the most concise, and focused, user experience should be preferred, with the HealthSonar app being, intentionally, a reporting tool.

E. System's data storage and processing

The radar data acquired during the HealthSonar system's operation are either stored and processed locally on the radar-based monitoring device, or uploaded to the cloud for storage and further processing. Whether the storage and processing is done offline or online is determined by how time-critical the resulting analytics are for the user.

The gait monitoring and, crucially, the fall detection pipelines should provide their outputs as close to real-time as possible. Nurses and physicians should receive the results of a TUG test when it concludes in order to plan ahead based on the mobility state of the test's subject. The results of a TUG test can indicate a high risk of falls in the future, thus such information is necessary in order to take preventative steps and arrange medical interventions before any misadventure takes place. Moreover, nurses and caregivers should receive notifications for a fall immediately in order to take timely action. Falls can lead to serious injuries, and given that they can go unnoticed if the faller is unable to move, real-time notifications can be a matter of life and death. As a result, all processing for those tasks is done locally on the radar-based device itself in order to avoid the time-consuming data uploading procedure or any Internet connection issues that may lead to problems in data uploading altogether. Nevertheless, as the TUG test raw radar data can be useful for research purposes, or simply for further custom analysis, there is the option to upload them to the cloud for storage and future use. This can be done via the web dashboard application.

On the other hand, sleep monitoring, does not provide time-critical data to the users. Generally, nighttime monitors are meant to gather data while a person is asleep, conclude the recording when they wake up, followed by processing the stored data and presenting them in a report when available. Note that, although sleep apnea events identified by the HealthSonar system during a person's sleep can be serious, they are not time-critical information and there is no advantage in sending real-time notifications to alert either the person sleeping or nurses and caregivers. As a result, the sleep monitoring pipeline includes automatically uploading the stored data to the cloud, as soon as the user wakes up, where they will be further processed and presented in a sleep report via the mobile app or the web dashboard.

Last but not least, presence detection is a necessary step prior to sleep monitoring and fall detection. As a process, it does need to run in real-time but it does not generate any data for further processing, apart from an event indicating that the sleep or fall pipeline should initiate. Hence, presence detection has to run locally on the HealthSonar device but no data are stored (for long-term use) or uploaded to the cloud.

F. Application programming interface (API)

The communication between the cloud services and the "local" parts of the HealthSonar system is enabled by an application programming interface (API). The API acts as the middleman for data access and integration between databases, web services and the device itself, as well as the mobile application. More specifically, the API is accessed through the web dashboard and the mobile application with the goal of bringing into contact the device with the cloud. The functionality offered by the API serves a pivotal role in the use of the HealthSonar and enables building its overall ecosystem. The facilitated communication is conducted in a secure manner to safekeep the sensitive user data.

G. HealthSonar: Customizable, extensible

The HealthSonar system was built to be customizable and extensible. As a result, the system can be used as a single

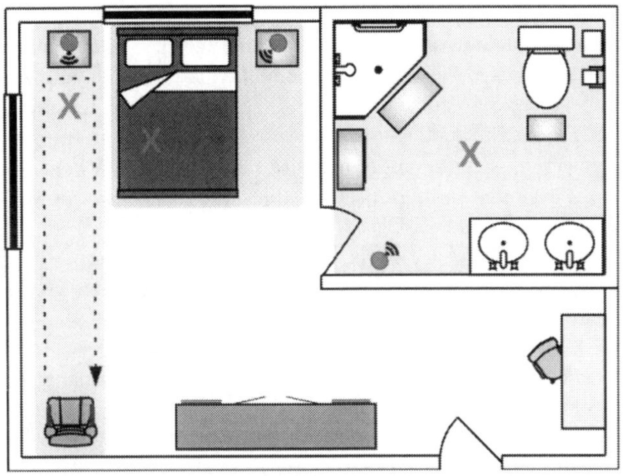

Fig. 3. Three suggested scenarios with the HealthSonar system inside a typical bedroom with a bathroom. **Green area**: A falling detection scenario. **Red area**: A sleep monitoring scenario. **Blue area**: A suggested Timed Up and Go (TUG) test scenario.

unit or as a "network" of radar-based devices, with all the radar's parameters able to be finely-tuned depending on the application at hand. The term "network" does not denote that the devices are connected with each other, but with a service offering centralized management of all the active HealthSonar devices of the organization using them. The centralized management is conducted via the web dashboard application.

The HealthSonar system is customizable enough to offer a plethora of options regarding how to use it and which features are needed for a certain application. More specifically, one can use at least one radar-based device along with all its accompanying software (namely, web portal, web dashboard, mobile app), or only with the ones they need. For research purposes, the web portal could offer complete functionality and elevated access to the device's embedded radar parameters, although the administrator should be knowledgeable regarding the theory and operation of IR-UWB radars.

On top of that, HealthSonar is extensible and can be deployed with one, or with as many radar-based devices is necessary. The extensibility is particularly important for care homes, and generally for clinical environments, as their needs change over time based on the number, and needs, of their residents or inpatients. The ease of extending, or even shrinking, the "network" of connected HealthSonar devices, can also help the organizations utilizing it to better-manage their human resources easily based on their needs. In a moment's notice, routine, yet important monitoring tasks, could be easily arranged to be performed by the HealthSonar system itself, leaving the personnel free to provide other vital tasks.

III. Presence, Sleep, Mobility and Falls

The HealthSonar system uses the IR-UWB radar module in order to capture rich micro and macro body movements that are subsequently used for monitoring health through: (1) detecting human presence within a space (2) studying nighttime sleep (staging, duration), (3) identifying sleep-related events (sleep apnea), (4) evaluating mobility (gait evaluation), and (5) detecting sudden, dangerous fall events. A description regarding the applications of the HealthSonar system (presence detection, sleep monitoring, gait evaluation and fall detection) can be found in more detail in the subsequent Sections III-A, III-B, III-C and III-D.

The performance of the HealthSonar system in those applications was evaluated, in a sleep lab of Evangelismos General Hospital (ethical committee's approval number 198/06-06-2022), with patients suffering from sleep disorders, in the neurology clinic of the University Hospital of Ioannina (ethical committee's approval number 7/21-03-2023), with patients suffering from movement disorders, as well as in-house, with healthy individuals, through experimental setups similar to the ones depicted in Fig. 3.

A. Presence detection

Presence detection is the cornerstone of the HealthSonar's features as it is a necessary step taking place prior to both sleep monitoring and fall detection. Presence detection is a procedure running constantly (apart from when the device is used for mobility evaluation), by default, on the HealthSonar device, as it is a prerequisite for initiating either the sleep monitoring or the fall detection pipeline.

Through presence detection, the HealthSonar system is capable of detecting whether someone is lying on a bed with the intention of sleeping, resulting in automatically initiating the sleep monitoring pipeline, based on the absence of activity and the monitoring of vital signs. On top of that, presence detection enables the bathroom wall-mounted HealthSonar device to identify whether someone is using the bathroom, resulting in automatically initiating the fall detection pipeline based on abrupt decreasing changes in signal's energy.

B. Sleep monitoring

Sleep monitoring is an important aspect of the HealthSonar system's functionality. Monitoring sleep in enabled for a nightstand-mounted HealthSonar device by the presence detection pipeline. More specifically, after the system detects a person lying on the bed, and until they move away from it, the sleep monitoring pipeline is initiated and runs continuously gathering data. Once the person moves away from the bed, the sleep monitoring routine comes to a halt and the generated data, which up to this point were stored locally on the HealthSonar device, are uploaded to the cloud for further processing, leading to the extraction of rich sleep-related metrics pertaining to sleep staging classification (including wake/sleep identification) and sleep apnea events detection that are presented either through the web dashboard or the mobile application. Contrary to the fall detection and gait evaluation pipelines, the processing of the gathered sleep data takes place on the cloud as there is no imminent need for the user to receive the extracted nighttime metrics in real-time. It is important to note that, during the sleep monitoring routine, the respiratory rate and the heart rate are calculated as they are a necessary preprocessing step prior to sleep staging and sleep apnea detection.

C. Gait evaluation

The HealthSonar system evaluates the gait of an individual based on the Timed Up and Go Test (TUG) [8], a well-established, standardized test, used for assessing various aspects of one's mobility, such as gait, balance and risk of falling [9], [10]. Performing a "classic" TUG test requires minimal instrumentation, namely an armchair (it is important for the chair to have arms), a mark for the 3 m distance and a stopwatch, although there are versions of the TUG test where extra instrumentation is used to better track the performance of the participants, (iTUG tests) [9]. A participant starts the test while sitting on the armchair, then, after being instructed, proceeds to stand up unaided, walk for 3 m, turns around the mark 180°, walks back to the armchair and sits down, thus, ending the test. The performance of the participant is timed with the stopwatch, and the total time of the test is used as the final score.

In order to evaluate gait using the HealthSonar device, a modified version of the TUG test is utilized. The modification lies in the use of a 3 to 5 m walking path instead of the "classic" 3 m, as it leads to the acquisition of more gait data, and as a result to a richer representation of the participant's mobility state. A suggested setup for a TUG test scenario with the HealthSonar can be seen in Fig. 3, where the device placed on top of the nightstand can be positioned in such a way as to monitor the path in front of the armchair. Note that due to the use of the radar-based device, this modified TUG test is considered instrumented (iTUG). The test results in the generation of a body of rich gait data, which are used to extract a set of gait metrics useful for the evaluation of a person's mobility. Those are, the total walking duration, the turning duration and the average gait speed [11].

The gait evaluation pipeline of the HealthSonar device does not run automatically, but is initiated manually through the mobile app (or the web portal to the radar for advanced, or custom use cases) by arranging a TUG test session. After the TUG test is finished, the resulting data are stored and processed locally on the HealthSonar device itself and the gait evaluation metrics are presented to the user via the mobile app through a Bluetooth connection. Note that there is the option to manually upload the generated TUG test data to the cloud storage though the web dashboard for future use.

D. Fall detection

Fall detection is a pipeline initiated through the presence detection feature of the bathroom wall-mounted HealthSonar device. The detection of fall events was designed to run for the bathroom area as it constitutes one of the most hazardous areas for elders and patients [12], [13]. Due to its nature, the fall detection pipeline is run locally, and continually (in case the presence detection procedure indicates there is a person using the bathroom), on the HealthSonar device as it needs to monitor for falls in real time. In the event of a fall, the system will identify it and sent a notification to both the web dashboard, as well as to the mobile phone application in order to notify caregivers to attend to the faller's needs. Fall detection is a crucial feature for any system facilitating assisted living of elders and patients as falls can result in serious, even fatal, injuries, thus their identification in real time can lead to timely interventions preventing further serious consequences for the faller.

IV. Discussion

In this work, a complete, unified, assisted living solution, the HealthSonar, targeted at monitoring sleep, evaluating gait and identifying falls was presented. The innovative nature of the solution lie in the fact that it is cheap, easy to use and home-based, while offering all the aforementioned features combined. As far as we know, currently no other alternatives exist that tick all those "boxes". The HealthSonar system was designed and built in-house, from the ground up, into a fully functioning assisted living monitor. The system was designed to be operated with minimal to no interaction from the user. Apart from the installation and the first setup, it was of paramount importance for HealthSonar to be unobtrusive while continually monitoring its surroundings, either for sleeping individuals or falls. As a matter of fact, it is possible for the user to bypass using any of its software application and "ignore" the system altogether. In that case, a treating physician or a nurse can be responsible for reading reports about the user's sleep, as well as receive notifications regarding falls. It is important to note that the device has no battery and can be permanently connected to a wall socket, thus existing on top of the nightstand (or the wall) indefinitely.

As of now, the functionality of the HealthSonar system is significant, yet future directions could extend it in various ways. One of the most useful, lie in the improvement of the mobility evaluation to include the identification of freezing of gait events in patients with Parkinson's disease. Moreover, as already mentioned, fall detection within the bedroom area could also be explored in the future based on the feedback we will gather from more users of the system. Last but not least, the heartbeat extraction, being a preprocessing step for sleep monitoring, could be further improved. Last but not least, the device was built as a prototype, thus technical limitations prohibited the design and construction of a more refined case for the inner hardware. As a result, in a future version, the device could be redesigned from the ground up, and one particularly useful addition would be a screen showing the time, the weather or the news of the day. This feature would be useful for the nightstand-mounted device, as it would then serve as a dashboard showing useful information.

V. Conclusions

The population of elders and patients with sleep and neurological disorders is rising, leading to a growing need of accurate continuous health monitoring, mainly revolving around sleep and mobility. Continuous, privacy-preserving, contactless, monitoring solutions, such as the HealthSonar, have the potential to disrupt the way clinical organizations (care homes, clinics, hospitals, etc.) monitor their residents, by providing a stream of health analytics, while helping them preserve human resources from repetitive, yet necessary tasks such as continuous health monitoring.

Acknowledgment

This research was funded by the European Regional Development Fund of the European Union and Greek national funds through the Operational Program Competitiveness, Entrepreneurship, and Innovation, under the call RESEARCH–CREATE–INNOVATE (project name: HealthSonar, project code: T2EDK-04366)

References

[1] D. T. Kasai, "Preparing for population ageing in the Western Pacific Region," *The Lancet Regional Health - Western Pacific*, vol. 6, p. 100069, Jan. 2021.

[2] J. Acquavella, R. Mehra, M. Bron, J. M.-H. Suomi, and G. P. Hess, "Prevalence of narcolepsy and other sleep disorders and frequency of diagnostic tests from 2013–2016 in insured patients actively seeking care," *Journal of Clinical Sleep Medicine*, vol. 16, no. 8, pp. 1255–1263, Aug. 2020.

[3] N. P. Gordon, J. H. Yao, L. A. Brickner, and J. C. Lo, "Prevalence of sleep-related problems and risks in a community-dwelling older adult population: a cross-sectional survey-based study," *BMC Public Health*, vol. 22, no. 1, p. 2045, Nov. 2022.

[4] C. Ding, Y. Wu, X. Chen, Y. Chen, Z. Wu, Z. Lin, D. Kang, W. Fang, and F. Chen, "Global, regional, and national burden and attributable risk factors of neurological disorders: The Global Burden of Disease study 1990–2019," *Frontiers in Public Health*, vol. 10, p. 952161, Nov. 2022.

[5] G. Tiberi and M. Ghavami, "Ultra-Wideband (UWB) Systems in Biomedical Sensing," *Sensors*, vol. 22, no. 12, p. 4403, Jun. 2022.

[6] D. N. Wickramarachchi, S. P. Rana, M. Ghavami, and S. Dudley, "Comparison of ir-uwb radar soc for non-contact biomedical application," in *2023 IEEE 17th International Symposium on Medical Information and Communication Technology (ISMICT)*. IEEE, 2023, pp. 01–06.

[7] D. K.-H. Lai, Z.-H. Yu, T. Y.-N. Leung, H.-J. Lim, A. Y.-C. Tam, B. P.-H. So, Y.-J. Mao, D. S. K. Cheung, D. W.-C. Wong, and J. C.-W. Cheung, "Vision Transformers (ViT) for Blanket-Penetrating Sleep Posture Recognition Using a Triple Ultra-Wideband (UWB) Radar System," *Sensors*, vol. 23, no. 5, p. 2475, Feb. 2023.

[8] D. Podsiadlo and S. Richardson, "The Timed "Up & Go": A Test of Basic Functional Mobility for Frail Elderly Persons," *Journal of the American Geriatrics Society*, vol. 39, no. 2, pp. 142–148, Feb. 1991.

[9] P. Ortega-Bastidas, B. Gómez, P. Aqueveque, S. Luarte-Martínez, and R. Cano-de-la Cuerda, "Instrumented Timed Up and Go Test (iTUG)—More Than Assessing Time to Predict Falls: A Systematic Review," *Sensors*, vol. 23, no. 7, p. 3426, Mar. 2023.

[10] B. M. Kear, T. P. Guck, and A. L. McGaha, "Timed Up and Go (TUG) Test: Normative Reference Values for Ages 20 to 59 Years and Relationships With Physical and Mental Health Risk Factors," *Journal of Primary Care & Community Health*, vol. 8, no. 1, pp. 9–13, Jan. 2017.

[11] A. Ntanis, N. Kostikis, I. Tsimperis, K. Tsiouris, G. Rigas, and D. Fotiadis, "Evaluating Parameters of the TUG Test Based on Data from IMU and UWB Sensors," in *2022 18th International Conference on Wireless and Mobile Computing, Networking and Communications (WiMob)*. Thessaloniki, Greece: IEEE, Oct. 2022, pp. 142–147.

[12] R. Blanchet and N. Edwards, "A need to improve the assessment of environmental hazards for falls on stairs and in bathrooms: results of a scoping review," *BMC Geriatrics*, vol. 18, no. 1, p. 272, Dec. 2018.

[13] J. A. Stevens, E. N. Haas, and T. Haileyesus, "Nonfatal bathroom injuries among persons aged $\geq$ 15years—United States, 2008," *Journal of Safety Research*, vol. 42, no. 4, pp. 311–315, Aug. 2011.

Can the IR-UWB Radar Sensor Substitute the PSG-Based Primary Vital Signs' Measurements?

Anastasia Pentari[1], George Rigas[2], Adamantios Ntanis[2], Thomas Kassiotis[1], Dimitrios Manousos[1], Evangelia Florou[3], Emmanouil Vagiakis[3], Dimitrios Fotiadis[4] and Manolis Tsiknakis[1,5]

[1]Institute of Computer Science, Foundation for Research and Technology—Hellas, Heraklion, Greece, GR 700 13
[2]PD Neurotechnology Ltd., Ioannina, Greece, GR 455 00
[3]Sleep Disorders Center, Intensive Care Unit, Evangelismos Hospital, Athens, Greece, GR 106 76
[4]Biomedical Research Institute, Foundation for Research and Technology—Hellas, Ioannina, Greece, GR 455 00
[5]Department of Electrical and Computer Engineering, Hellenic Mediterranean University, Heraklion, Greece, GR 710 04

Corresponding author: Anastasia Pentari, e-mail: anpentari@ics.forth.gr

Abstract—**During the last decade, the development of impulse radio ultra-wideband (IR-UWB) radar sensors have led them to be considered as a viable substitute of polysomnography (PSG), the gold standard, in the acquisition of the primary vital signs of the human body, during sleep. In this work, we investigate whether radar sensor recordings of the chest and heart displacement can accurately substitute the PSG chest and heartbeat signal measurements. We develop an innovative pipeline of handling the radar-based recordings, which includes: motion detection, extraction of the respiration, heartbeat and activity vital signs and estimation of the respiratory and heartbeat rates (RR and HR, respectively). Next, we apply our proposed methodology to data from 28 subjects gathered during their sleep. Results show that the radar sensor's measurements can be comparative to those produced by the PSG. Specifically, the RR and HR frequencies, of the radar and the PSG, have average Pearson's correlation, greater than** 0.9 **and** 0.8**, respectively.**

Index Terms—**IR-UWB radar sensor, PSG, respiratory rate, heartbeat rate, activity signal**

I. Introduction

Conducting a sleep study is the most important procedure for the identification and assessment of various sleep disorders, such as obstructive sleep apnea (OSA), a serious medical condition, that can even result in cognitive dysfunctions, cardiovascular and cerebrovascular diseases [1]. The most frequently-used diagnostic tool is polysomnography (PSG), which is widely considered the gold standard [2]. Nonetheless, the multiple on-body sensors, required for it, increase the patients' discomfort, while its long duration renders it expensive and appropriate only for use in hospitals and research settings [3].

To remedy the problems encountered during the use of the PSG, impulse-radio ultra-wideband (IR-UWB) radars have started to be extensively used during sleep studies. These sensors allow for contactless monitoring and have the ability to record the motion of the body, ranging from the micro-motion of the chest to the macro-motion of objects, or people, moving with high accuracy [4], [5].

Radar-based devices are convenient tools for sleep monitoring at home settings, as they are user-friendly and can accurately identify chest and heart oscillations. Those generated data can be subsequently used for extracting the respiratory rate, the heartbeat and the activity signal (that is the activity of the human body during sleep) which provide crucial information about sleep quality [6].

Radars are sensitive to environmental noise and the rich generated body motion information can be considered as "noise" in the context of sleep monitoring. As a result, a pipeline for noise/motion detection is used for the extraction of more accurate respiration and heartbeat signals. Specifically, the radar sensor's recordings are analyzed through a spectrogram-based approach. Spectrograms are capable of providing useful information about the body motion, through the changes of radar's characteristics over time [7]. In essence, through the spectrogram-based analysis light and intense motion can be detected and removed. To that end, the activity signal is also used, as it provides motion information that enables the researchers to exclude parts of the recordings that contain intense noise and extract, more reliably and accurately, the respiratory and heartbeat rates (RR and HR, respectively). This is an important point for consideration, given the impact that those frequencies have for a sleep study.

In this work, we investigate whether the vital signs extracted from a radar sensor are a reliable representation of those acquired from a PSG. Specifically, the chest and electrocardiogram signals acquired from the PSG are compared with those acquired from the radar by estimating the corresponding RR and HR frequencies. Moreover, from the radar data the activity signal is extracted that is used for motion detection.

The main contributions of this study are:

1) The creation of a system for the analysis of radar-based recordings which can provide three signals, namely the respiration, the heart displacement and the activity.
2) The noise treatment of the signals, that is introduced due to either body motion or environmental factors, by an innovative spectrogram-based analysis.

Simple, yet state-of-the-art, procedures are employed to provide the accurate extraction of biological signals (respiration

TABLE I
ARIA SENSING LT102 RADAR SPECIFICATIONS.

General specifications	Values
Radar's operating frequency	6.5 GHz to 8.5 GHz
Temperature operating range	−40 °C to 85 °C
Radar module's dimensions	36 mmx68 mm
Maximum power consumption	220 mW at 5 V
Integrated antenna aperture	±60° by ±60°
Typical detection range	12 m

and heart displacement) and subsequently the RR and HR. The evaluation of the pipeline was conducted mainly using the Pearson's correlation, between the PSG- and the radar-extracted results. Note that the evaluation was conducted using 28 recordings of patients with OSA disorder.

The rest of the paper is organized as follows: Section II describes the mathematical background of the proposed methodology, Section III, presents the evaluation of the method, while Section IV discusses the conclusions.

II. MATERIALS AND METHODS

In this section the main building blocks of the proposed methodology are described.

A. IR-UWB Data Acquisition

In our experiments, an IR-UWB radar sensor was positioned next to a patient, about $50cm$ from their chest, and captured the motion of their lungs and heart. The sensor used was the Aria Sensing LT102 and its specification are presented in Table I. Generally, the procedure of capturing biological signals using a radar, involves measuring the torso displacement caused by the motion of the lungs and the heart. IR-UWB radars track the amplitude of a human body point to extract biological signals [8], by detecting and quantifying the periodic expansion of the chest. Thus, an important parameter in the extraction of those signals is the distance between the radar antenna and the human chest, which changes over time. The distance can be described by the following equation:

$$d(t) = d_0 + a_r \sin(2\pi f_r t) + a_h \sin(2\pi f_h t) \quad (1)$$

where d_0 is the nominal distance and a_r, a_h are the mean values over the range of all possible displacements of the chest cavity caused by the respiration and the heart displacement, respectively. Moreover, with f_r, f_h we denote the respiratory and heartbeat frequencies, respectively [9].

The IR-UWB radar sensor's acquired data consist of a 2-dimensional matrix, that is a function of the captured samples K and the radar's bins M (equal to 359 in our case). In more detail, a bin is a unit of representing and organizing the radar's captured spatio-temporal information. The total number of bins is related to the distance between the human body and the radar sensor [9]. Specifically, the radar's output can be expressed as $\mathbf{S} \in \mathrm{R}^{\mathrm{K} \times \mathrm{M}}$, where K denotes the number of samples in the sample space and M the radar's bins in fast time (in nsec).

B. Preprocessing of the Radar Recordings

As mentioned earlier, the radar's recordings belong to a 2-dimensional space, with the bins being of major importance. Between consecutive recordings, the corresponding bins have stored almost the same human body information, with the difference that each recording differs from the other usually in the amplitude of the captured signals, i.e., the magnitude of the acquired displacements. As a result, the bin with the most prominent displacement should be identified. To that end, the most well-established and simple method of finding the bin of interest is by computing the variance of each bin between all recordings and then take the bin with the maximum variance [9]. In our experimental procedure, as some patients changed position quite often, we split the matrix into N non-overlapping segments, of 120 seconds duration, and for each segment we repeated the aforementioned procedure. After that, by combining the signal's parts with the most intense human chest and heart displacement, we constructed a new signal of interest, denoted as $s \in \mathrm{R}^{1 \times \mathrm{K}}$.

C. Extraction of the Activity Signal

The activity signal is among the most important signals in the context of sleep disorders. The presented methodology of detecting motion was based on the computation of the spectrogram of the extracted signal, denoted as s. Specifically, the procedure depends on the algorithm described in Algorithm 1. Inputs to the algorithm are the signal extracted from the radar-based recordings, namely s, as well as the sampling frequency of the radar, which in our case was equal to 40Hz.

Next, the spectrogram of each examination is estimated. A spectrogram is presented in Fig. 2. The algorithm detects low frequency time points, i.e., time points which present non-uniform behavior indicating the presence of a "light-tailed" distribution of motion within the signal (taking place during the time when the person was awake, yet lying in a bed). Note that the majority of time points do not include light motion i.e., they were captured during sleep. Moreover, given that the recordings took place during a sleep examination, in which people were lying in a bed, intense motion would not be present. To identify light (i.e., rare) motion (the signal samples that are presented with dark blue color in the spectrogram of Fig. 2), a threshold of 0.3 multiplied by the maximum value of the spectrogram matrix (line 9 of Algorithm 1) was applied. Having identified the areas within the signal where light motion existed, the signal was converted into a binary form by replacing the value of points where light motion existed with a value of 1, otherwise with a value of zero. To do so, a threshold denoted as $thresh$, equal to 8 or 10, was applied. Regarding the threshold value, it was important to not be excessive leading to the loss of valuable information, yet not as small as to eliminate the signal's fluctuations which could be important in differentiating patients' sleep apnea events, for instance.

The proposed methodology proved to be robust, although the noise was not intense. In Fig. 3, an example of an activity signal is presented. The proposed methodology has captured,

Inputs: s, F_s
Outputs: x
Spectrogram Computation: $\mathbf{S_p} = spectrogram(s, F_s)$
$\mathbf{S_p} = 10\log_{10}(\mathbf{S_p})$
Estimation of maximum spectrogram's value (in dB): $maxV = \max\left[\mathbf{S_p}(:)\right]$
Define: $N_t : length\left[\mathbf{S_p}(:,1)\right]$, $N_f : length\left[\mathbf{S_p}(1,:)\right]$
for t = 1:N_t **do**
  for f = 1:N_f **do**
    if $\mathbf{S_p}(t,f) \leq 0.3 \cdot maxV$ **then**
      $\mathbf{S_b}(t,f) = 1$
    end if
  end for
end for
Define: $L = length(s)$
Define: $step = L/N_f$
Define: $freqs = [1 : step : L - step]$
Define: $counter = 1$
for f = 1:N_f **do**
  $part = \mathbf{S_b}(:,f)$
  $c = \sum(part == 1)$
  if $c \geq thresh$ **then**
    $\mathbf{S_b}(t,f) = 1$
    for i = $freqs(counter) : freqs(counter+1)$ **do**
      $x(i) = 1$
    end for
  end if
  $counter = counter + 1$
end for

Fig. 1. Activity signal extraction procedure.

not only the intense signal's noise (probably from motion), but also the lower amplitude noise (probably from the environment). As a result, it is evident that through the appropriate analysis of the spectrogram, the detection of human motions, both during sleep and during daytime activities, is possible [7].

D. Extraction of the Respiration Signal and the Respiratory Rate

The respiration signal is of major importance for the study of sleep and especially for the identification of sleep apneas. Thus, its accurate extraction from the radar-based recordings will result to a more accurate estimation of the RR. In our procedure, we followed the "maximum variance" process, which enables the detection of the signal's fluctuations with the maximum amplitudes. Our purpose was to estimate the respiratory rates with a simple, fast and accurate manner. The estimation of the RR was based on the computation of the power spectral density. Specifically, the first step was to split the previously extracted signal, s, into non-overlapping windows of 60 or 120 seconds duration, in order to maintain sufficient amounts of the signal's information. For each segment, if the corresponding activity signal's segment was full of zeros, the Fast Fourier Transform (FFT) was applied and then the power spectral density (PSD) was estimated. In the literature, the number of respirations during sleep range from 12 to 20 [10], which equals to a respiration frequency between 0.2 and 0.35 Hz. As a consequence, having estimated the PSD we search for its maximum value into the aforementioned frequency range. Then, we multiply the estimated frequency with 60 in order to convert it into a respiration frequency (i.e., respirations per minute). On the other hand, in the case that the activity signal had intense motion within a segment, a linear

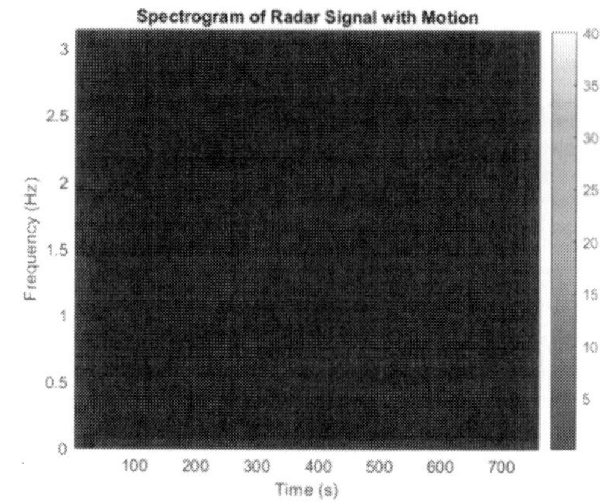

Fig. 2. An example of a spectrogram computed from patient data after a sleep monitoring procedure.

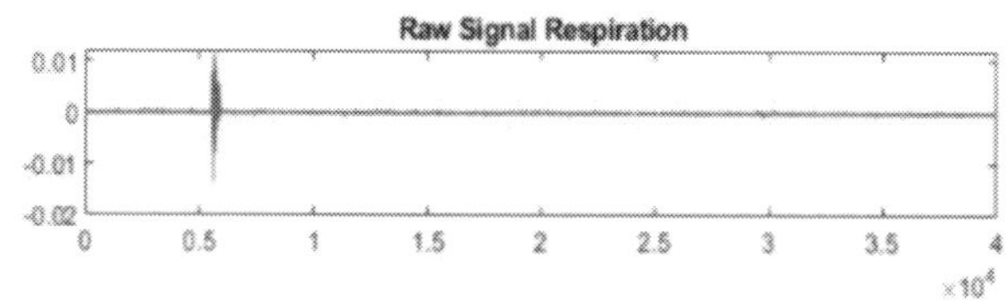

(a)

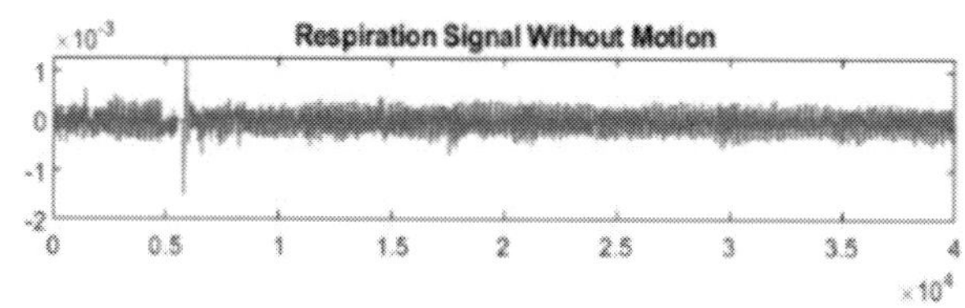

(b)

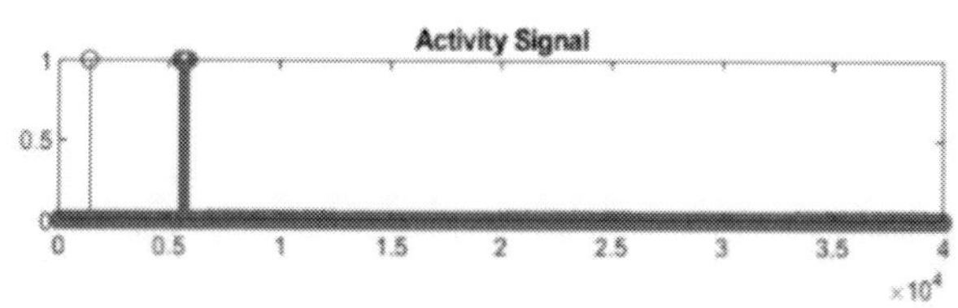

(c)

Fig. 3. The procedure for extracting an activity signal. (3a) Raw respiration signal from the radar recordings; (3b) The same signal'a part with reduced noise; (3c) the corresponding activity signal.

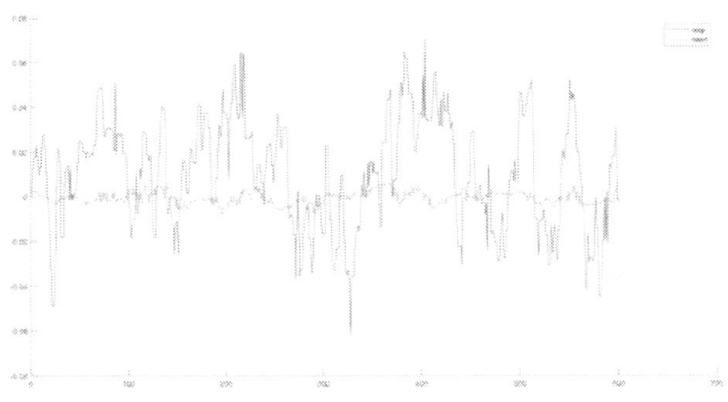

Fig. 4. Example of a radar's extracted signal. Black color denotes the respiration while red the heartbeat.

interpolation was applied that took into consideration the 2 immediate neighboring samples (both left and right).

E. Extraction of the Heartbeat Signal and the Heartbeat Rates

The heart signal is more laborious to be extracted from data of the radar-based device, as it is superimposed on top of the "stronger" respiration signal, as we can see in Fig. 4. Based on [11], the first step to reach to an accurate heartbeat signal extraction is to remove the fluctuations corresponding to the respiration frequencies. Thus, the respiratory signal extracted from the previous procedure was passed through a low-pass filter, with a cut-off frequency equal to 0.9Hz, and then through a high-pass filter of a cut-off frequency equal to 0.5Hz [10]. Finally, a median filter of order 20 was applied, for removing small signal's fluctuations, which are usually created by the system's noise (i.e., artifacts).

Having retained the most prominent fluctuations of the raw radar signal, the majority of the heartbeats have been retained as well. The final step is to estimate the corresponding HR values. Similarly to the RR estimation, for each heart displacement signal segment, if the corresponding segment of the activity signal is full of zeros, we apply peak detection. The HR is defined by counting the peaks per segment. On the other hand, in the case of a non-zero activity signal segment, again, the corresponding samples' rates were replaced with a linear interpolation that took into consideration the 2 immediate neighboring rates (both left and right).

III. Results

In this section we present the evaluation of our pipeline, based on the Pearson's correlation.

A. HealthSonar Protocol – PSG Dataset

Data were collected during the HealthSonar clinical study, a one-arm observational clinical study carried out at the Sleep Disorders Center of the Intensive Care Unit located within Evangelismos Hospital at Athens, Greece. The purpose of this study was to evaluate the accuracy and validity of identifying sleep stages and apnea-related events, using data from an unobtrusive, contactless monitoring device based on an IR-UWB radar sensor compared to the same metrics as produced by data from PSG (considered the gold standard). The study was approved by the ethical committee of Evangelismos Hospital (Reference No. 198, 6/6/2022). In order to be enrolled to the study, the participants needed to be above 18 years old and able to consent to the aspects of the study before filling an informed consent form. All participants completed a clinical evaluation and underwent an attended polysomnography session. The examinations followed the "Manual for the Scoring of Sleep and Associated Events" (v2.6) of the American Academy of Sleep Medicine. Each subject was monitored for an entire night, while their sleep stages were annotated at 30-second intervals. Demographic information of the participants are presented in Table II.

TABLE II
DEMOGRAPHIC INFORMATION OF THE STUDY PARTICIPANTS.

No. of subjects	28
Age range	23-68 years
Weight range	55-155 kg

The PSG data used for our analysis included the following sensors' signals:

1) the "chest" signal, responsible for recording the patient's respiration. Its sampling frequency was equal to 32 Hz.
2) the "ECG_{LA}" and the "ECG_{RA}" signals responsible for acquiring the heart beats. The sampling frequency was equal to 256 Hz. The final electrocardiogram (ECG) signal was computed by the formula:

$$ECG = ECG_{LA} - ECG_{RA} \quad (2)$$

It is worth mentioning that the signals, from both PSG and radar, were synchronized and the radar-based activity signal was also used in the analysis of the PSG recordings, in order to clean the signals from the patients' motion while improving the synchronization.

B. PSG-based RR and HR

In order to estimate the corresponding RR and HR frequencies from the chest and ECG signals, we followed the procedure described on Section II. Briefly, for the estimation of the RR, the chest signal recorded from PSG was split in 120-second windows and for each window the PSD was estimated based on the FFT transform. The most prominent frequency of the PSD was considered as the respiratory rate for each window. Regarding the HR, again, the signal was split in windows of similar duration and through a peak detection process, the peaks of the ECG, having amplitude greater than a specific threshold, were identified. This threshold was defined in the range of $[0.4 - 0.6]$ of the maximum signal's amplitude.

For the estimation of both HR and RR, the regions of activity (as described in Section II-C) were fitted with a linear curve produced by linear interpolation, which took into consideration the 2 immediate neighbors (left and right).

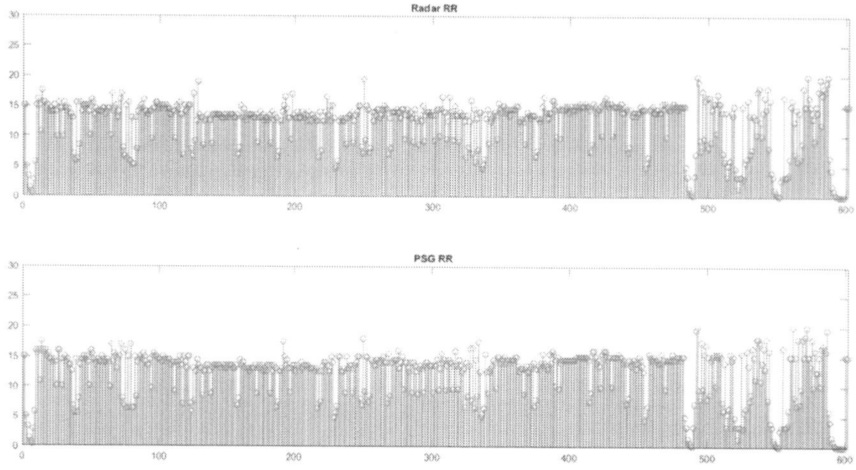

Fig. 5. The RR values for 5 randomly selected patients during their monitoring procedure as produced by the radar-based device (top) and the PSG system (bottom). The window duration is 120 seconds.

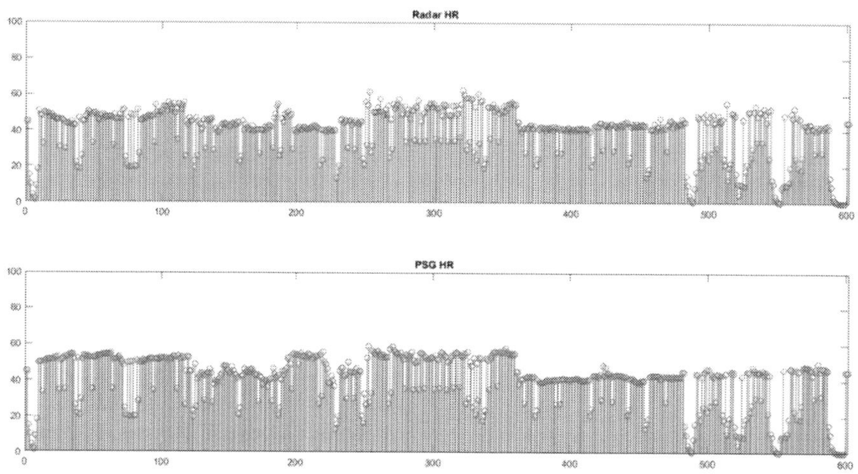

Fig. 6. The HR values for 5 randomly selected patients during their monitoring procedure as produced by the radar-based device (top) and the PSG system (bottom).

C. Performance Evaluation

In our analysis, the results were evaluated through 2 main metrics, i.e., the Pearson's correlation and the intraclass correlation coefficient (ICC) between the RR and HR measurements derived from the IR-UWB radar sensor and the measurements derived from the PSG system, as it was also used in [1]. The evaluation results can be seen in Table III. Note that the analysis was performed using windows with a duration of 120 seconds. Regarding the ICC values, the average correlation between the RR and the HR, as was extracted from the data radar and those of the PSG, was equal to 0.98 and 0.89, respectively.

In Fig. 5 and 6, we observe the RR and HR values for 5 selected patients during their monitoring procedure. Each spike corresponds to a rate derived from the analysis of a window of 120 seconds duration. As we can observe, especially for the case of the RR, the values estimated from data of both systems are very close, as it is verified by the high Pearson's correlation values and the ICCs. Notice that all raw frequencies were converted to represent the number of respirations and heart beats, respectively.

It is worth mentioning that after changing the window's duration to 60 seconds, the correlation values were lower and almost reached 0.8. Although still a high correlation, this result implies that our methodology has as a limitation the samples

TABLE III
PEARSON'S CORRELATION OF RR AND HR RATES BETWEEN THE RADAR SENSOR AND THE PSG SYSTEM.

Patient ID	Respiratory	Heartbeat
1	0.99	0.88
2	0.95	0.91
3	0.96	0.98
4	0.97	0.71
5	0.93	0.81
6	0.95	0.96
7	0.98	0.95
8	0.99	0.67
9	0.93	0.83
10	0.98	0.89
11	0.93	0.90
12	0.99	0.97
13	0.98	0.46
14	0.99	0.69
15	0.99	0.99
16	0.94	0.97
17	0.97	0.98
18	0.91	0.96
19	0.95	0.97
20	0.95	0.98
21	0.94	0.96
22	0.93	0.97
23	0.96	0.88
24	0.79	0.95
25	0.91	0.95
26	0.88	0.93
27	0.98	0.98
28	0.99	0.99
Average	0.95	0.90

we take into consideration. As a result, the 60-seconds duration seems to not be able to provide sufficient information for the procedure to estimate the rates with more accuracy.

IV. CONCLUSIONS

The goal of this study was to investigate whether the radar sensor can accurately substitute PSG, especially in the case of respiratory rate and heartbeat estimation. The analysis was focused on constructing an innovative pipeline which takes into consideration the radar-based recordings, represented by a 2-dimensional matrix and results in the following clinical information:

1) the respiration and heart displacement signals,
2) the respiratory and the heart rate, as well as
3) the activity signal (activity of the human body).

Results prove that the presented simple and fast procedure for processing the radar-based data generates accurate respiration and heart displacement signals, that are highly correlated with the corresponding PSG signals, considered as the ground truth. Moreover, the procedure for extracting the activity signal, was shown to be effective, leading to an improved performance in extracting the aforementioned signals, as well as estimating the corresponding frequencies.

Our methodology was accurate for most of the patient cases that was tested on, highlighting that it is robust to different patients' characteristics and also to different sleep patterns. Overall, the methodology is simple and fast, and the data analysis does not require a large number of external parameters apart from the cut-off frequencies of the low-pass and high-pass filters utilized for the heart displacement signal extraction, as well as an appropriate threshold for the extraction of the activity signal. However, our methodology has to be further tested using complementary features which can provide more information about the detection of the OSA.

V. ACKNOWLEDGEMENTS

* This research was funded by the European Regional Development Fund of the European Union and Greek national funds through the Operational Program Competitiveness, Entrepreneurship, and Innovation, under the call RESEARCH-CREATE–INNOVATE (project name: HealthSonar, project code: T2EDK-04366) and from the European Union's Horizon 2020 research and innovation program under grant agreement No 101017331 (ODIN).

REFERENCES

[1] J. W. Choi, D. H. Kim, D. L. Koo, Y. Park, H. Nam, J. H. Lee, H. J. Kim, S.-N. Hong, G. Jang, S. Lim *et al.*, "Automated detection of sleep apnea-hypopnea events based on 60 ghz frequency-modulated continuous-wave radar using convolutional recurrent neural networks: A preliminary report of a prospective cohort study," *Sensors*, vol. 22, no. 19, p. 7177, 2022.

[2] J. V. Rundo and R. Downey III, "Polysomnography," *Handbook of clinical neurology*, vol. 160, pp. 381–392, 2019.

[3] H. S. A. Heglum, H. Kallestad, D. Vethe, K. Langsrud, T. Sand, and M. Engstrøm, "Distinguishing sleep from wake with a radar sensor: a contact-free real-time sleep monitor," *Sleep*, vol. 44, no. 8, p. zsab060, 2021.

[4] W. H. Lee, S. H. Kim, J. Y. Na, Y.-H. Lim, S. H. Cho, S. H. Cho, and H.-K. Park, "Non-contact sleep/wake monitoring using impulse-radio ultrawideband radar in neonates," *Frontiers in Pediatrics*, vol. 9, p. 782623, 2021.

[5] L. Ma, M. Liu, N. Wang, L. Wang, Y. Yang, and H. Wang, "Room-level fall detection based on ultra-wideband (uwb) monostatic radar and convolutional long short-term memory (lstm)," *Sensors*, vol. 20, no. 4, p. 1105, 2020.

[6] X. Zhang, X. Yang, Y. Ding, Y. Wang, J. Zhou, and L. Zhang, "Contactless simultaneous breathing and heart rate detections in physical activity using ir-uwb radars," *Sensors*, vol. 21, no. 16, p. 5503, 2021.

[7] S.-w. Kang, M.-h. Jang, and S. Lee, "Identification of human motion using radar sensor in an indoor environment," *Sensors*, vol. 21, no. 7, p. 2305, 2021.

[8] D. Wang, S. Yoo, and S. H. Cho, "Experimental comparison of ir-uwb radar and fmcw radar for vital signs," *Sensors*, vol. 20, no. 22, p. 6695, 2020.

[9] A. Pentari, D. Manousos, T. Kassiotis, G. Rigas, and M. Tsiknakis, "Respiration and heartbeat rates estimation using IR-UWB non-contact radar sensor recordings: A pre-clinical study," in *Workshop Proceedings of the EDBT/ICDT 2023 Joint Conference*, Ioannina, Greece, Mar. 2023.

[10] R. Avram, G. H. Tison, K. Aschbacher, P. Kuhar, E. Vittinghoff, M. Butzner, R. Runge, N. Wu, M. J. Pletcher, G. M. Marcus *et al.*, "Real-world heart rate norms in the health eheart study," *NPJ digital medicine*, vol. 2, no. 1, p. 58, 2019.

[11] F. Khan and S. H. Cho, "A detailed algorithm for vital sign monitoring of a stationary/non-stationary human through ir-uwb radar," *Sensors*, vol. 17, no. 2, p. 290, 2017.

Proposal and Evaluation of Optical Sensor to Identify Liquids in Liquid Intake Detection System Using Smart Bottles

Sandra Viciano-Tudela[1], Paula Navarro-Garcia[1], Lorena Parra[1], Sandra Sendra[1], Jaime Lloret[1*]
[1]Instituto de Investigación para la Gestión Integrada de Zonas Costeras, Universitat Politècnica de València, 46730 Grau de Gandia, Spain
Email: svictud@upv.es, paunagar@alumni.upv.es, loparbo@doctor.upv.es, sansenco@upv.es, jlloret@dcom.upv.es

Abstract— **Dehydration poses health risks, leading to confusion, falls, and even death. Tracking water intake, especially among at-risk groups, is vital. Smart bottles with sensors offer a solution for estimating and monitoring fluid consumption efficiently and widely. The paper aims to enhance the existing liquid intake detection systems by developing an optical sensing element to differentiate various liquids for precise hydration assessment. The study evaluates data processing techniques, including classic statistics, Discriminant Analysis, and Artificial Neural Networks, to classify liquids. The system is based on an ESP32 node integrated into the smart bottle as an Internet of Things device with communication capabilities with other wearable devices. A total of 7 different liquids are included in the conducted experiments. The data-gathering process is repeated several times to generate training and verification datasets. The results indicate that it is possible to differentiate the liquids using a reduced number of light wavelengths, white and purple. All analyzed techniques offered good results. Discriminant Analysis is the most effective classification approach with 100% accuracy. Nevertheless, if distinguishing between different types of teas is not necessary, thresholds based on statistical tools can be employed using fewer computation resources.**

Keywords-RGB sensor; photoreceptors; Discriminant Analysis; Artificial Neural Network; Internet of Things

I. INTRODUCTION

Dehydration is a very important health problem, which can cause: confusion, falls, hospitalization, and, in the worst cases, death. That is why it is important to keep track of daily water consumption, especially in risk groups, such as the elderly and people with diseases that affect fluid regulation [1]. Despite the fact that there are groups that have priority, citizens' awareness of water consumption has increased since it has been shown that staying well-hydrated is essential both for our physical well-being and for cognitive health [2]. A survey was carried out where it was observed that 70% of those surveyed under the age of 50 admitted they had forgotten to drink water or feared they had not done so due to their hectic lifestyle [3]. Being the consumption of water an issue that affects not only risk groups but also the rest of the population.

Regarding the current solutions for stimulating fluid intake, multiple proposals can be found based on technological systems. On the one hand, there are systems, mainly integrated into smart devices, such as smartphones or smartwatches, that indicate the necessity of drinking every certain given period of time [4]. On the one hand, there are systems that monitor drinking habits based on sensors embedded in wearable devices. Nevertheless, in this case, most of these systems are designed to identify problems linked with alcohol consumption [5]. The use of sensors embedded in a smart bottle that determines fluid intake is limited to medical surveillance [6], athletes [7], or elderly people [8]. For all this, it would be convenient to find a system that allows estimating daily water intake efficiently and simply with applications from hospitals and other health centres to the rest of the population.

Smart bottles are portable devices capable of detecting the type and amount of liquid you ingest, thanks to the use of sensors and physiological parameters [2]. Users can use and manage them in different ways: using a smartphone or uploading the data to the cloud [9]. Regarding the technologies used, the Liquid Intake Detection System (LIDS) provides real-time monitoring of the type and volume of fluid intake. For this, the system designs a detection module comprising ultrasonic, Red-Green-Blue (RGB) colour, temperature and accelerometer sensors, as well as a computational framework for classifying the type of fluid intake [2].

The aim of this paper is to develop the RGB sensing element of the LIDS, which is capable of distinguishing between different liquids. Among existing systems, smart bottles are focused on quantifying the liquids without distinguishing the type of liquid. Determining the sort of liquid is important in order to establish hydration needs. Therefore, the proposed system is tested with 7 different sorts of liquids in this paper. As the optical sensor, an RGB Light Emitting Diode (LED) module is used and configured to emit 7 different light colours and a photodetector. Calibration and verification tests are conducted to evaluate different data processing techniques to maximize the number of classified liquids. Classic statistics, Discriminant Analysis (DA), and Artificial Neural Networks (ANN) are compared among data processing techniques.

The rest of the paper is structured as follows: Section 2 outlines the related work. The proposal, including the used sensors, node, and architecture of the LIDS system, is described in Section 3. Section 4 details the test bench. The results of the gathered data are analyzed in Section 5. Finally, Section 6 summarises the conclusions and future work.

II. RELATED WORK

In 2016, a study was carried out comparing different ways of handling smart bottles. In this study, an intelligent water bottle was made with the aim of being able to monitor the consumption of water in a day produced by a person. In order to achieve this, they used sensors of inertia and physiological parameters and a photoplethysmographic sensor. The user could manage this bottle in two different ways: using a smartphone or uploading the data to the cloud. The use of a smartphone is more energy efficient, but you need to be constant during the day, uploading the data periodically. However, the cloud-based system was less energy efficient, but it helps structure many users' data and provides more reliable information [9].

In 2021, research began on the technologies that should be used to make them more efficient and practical for users.

In the first, the LIDS system was studied. In this article, they have focused on creating a system capable of characterizing the type of liquid consumed and its volume. The LIDS provides real-time monitoring of the type and volume of fluid intake. For this, the system designs a detection module comprising ultrasonic, RGB colour, temperature and accelerometer sensors, as well as a computational framework for classifying the type of fluid intake [2].

The second one exposes the available technologies, reviews the existing systems for monitoring the amount of water consumed by users and motivates them to do so by sending notices periodically during the day. It was concluded that the best results are obtained by combining the technologies that are available today: wearable devices, surfaces with integrated sensors, solutions based on vision and the environment, and smart containers [8].

Later, in 2022, the same authors did a study comparing the performance and functionality of four smart bottles on the market. The bottles that were compared were H2Opal, HidrateSpark Steel, HidrateSpark 3 and Thermos Smart Lid. To know the effectiveness of each model, 100 intakes for each bottle were recorded and analyzed, comparing the amount of water consumed using a high-resolution weight scale and the data obtained from the bottle. It was concluded that the best options were the first three, the Smart Lid thermos being the least effective [1]. In this same year, a smart bottle was used for sanitary purposes. The Hydrate Spark bottle was used to count the amount of water ingested by a user prone to stone formation. This bottle had sensors that counted the amount of water in real-time and sent that information to the smartphone, which tells you when you have drunk enough water according to your goal [10].

Among the surveyed proposals, no one of them has focused on developing a system that measures the type of liquid included in the smart bottle. As far as we are concerned, no proposal aimed at classifying the liquids inside the bottles used for LIDS in order to differentiate among intakes of liquids. Thus, this proposal aims to classify the included liquid based on the combination of different patterns of light abortion and include this data in machine-learning solutions, such as DA and ANN.

III. PROPOSAL

This section describes the optical sensor used to detect the different samples in the smart bottle. In addition, the node used and the system's complete architecture are shown.

A. Optical Sensor

This prototype consists of a tube 15 cm high. It is an opaque black PVC tube. In this way, the interference of outside light with the sample is avoided. On the one hand, an RGB LED has been used to differentiate the different test substances, allowing the Arduino programming to establish different wavelengths. In this case, 8 colours have been used: red, green, blue, yellow, purple, cyan and white. The following table, Table 1, shows the colours, intensity and RGB values.

TABLE I. RGB LED INTENSITIES FOR USED COLOURS

Light	Red	Green	Blue
Red	255	255	0
Green	255	0	255
Blue	255	0	0
Yellow	255	255	255
Purple	255	255	0
Cyan	255	0	255
White	255	255	255

On the other hand, the optical sensor contains an infrared LED. The RGB and infrared LED are arranged on the same side. On the opposite side are the visible and IR photoreceptors that make it possible to detect the light that passes through the sample. With this system, it is feasible to detect the absorbance of the tested samples and compare the data between them, which is used to differentiate among tested liquids. The system is based on an enhanced RGB prototype version presented in [11] and [12].

B. Node

An ESP32 microprocessor has been selected to gather and analyze the data. It has been chosen due to its small size, which allows it to be included in the smart bottles in a comfortable way for the user. It has good connectivity and is widely used for Internet of Things (IoT) devices. The different inputs it presents allow the RGB LED to be connected digitally and the Near Infra Red (NIR) LED analogically. The node is presented in Figure 1.

Figure 1. Picture of used node

C. Architecture

The architecture that is proposed for the LIDS-based smart bottle includes different networks. Firstly, the data obtained from the optical sensor are processed in the node itself by means of the ESP32 microcontroller as part of edge computing. After processing the data, the smart bottle will send the information obtained to a server with a database by connecting with the smartphone using the Bluetooth connection to reach the smartphone and then using the 5G network to reach the database in the server. We have selected using the smartphone and Bluetooth technology in order to ensure connectivity in indoor and outdoor environments regardless of local WiFi connectivity.

Moreover, through a smartphone, the user or the person's caregiver can access the data as a consult to the database. In this way, you can keep track of the liquids ingested by the person as well as the volumes.

When the person is at home, the smart bottle can be connected to a smartwatch or tablet by a WiFi connection. This makes it possible to establish an alarm system that indicates that the user must ingest liquids according to the time that has passed since the last ingestion.

IV. Test Bench

This section shows how the samples have been prepared for processing, in addition to how the data has been collected. Finally, the tools used for data processing and subsequent analysis are described.

A. *Sample preparation*

To develop the intelligent bottle, it used samples for different substances. The processing samples have been: empty, water, tea, Mango tea, light milk, milk, tea with milk and more tea with milk. To prepare the different teas, we introduced the tea to the water. After this, we mix by shaking. The samples are introduced into the intelligent bottle, which is the different colours of LEDs and IR. In the following table (Table 2), we show the characteristics of the samples and the volume included.

TABLE II. RGB LED INTENSITIES FOR USED COLOURS

ID	Sample	Volume (mL)
1	Empty	0
2	Water	50
3	Tea	50
4	Mango tea	50
5	Light milk	50
6	Milk	50
7	Tea with milk	50
8	More tea with milk	50

B. *Measuring process*

This subsection describes the measuring process. For data collection, the samples are introduced into the smart bottle. Using the SP32, the different colours of RGB LED are programs. We avoid collecting samples' values when the LEDs and IR illuminate samples. The data are saved like CSV, and after that, are processed.

Each sample is measured 6. In order to generate two datasets, training and validation, the data gathered are split into two groups. The first four replicas were used for the training dataset, while the last two replicas were used for verification.

C. *Data processing*

Three different approaches to classifying data as a traditional multi-class problem are compared in this paper. First, the ANalises Of Variance (ANOVA) is used as the most simple approach. A series of thresholds are generated based on the results of the multiple groups of the ANOVA. Then, DA and ANN are applied to compare the result of machine learning with simpler methods. The number of wrong classifications in both training and validation datasets is used as a performance indicator.

V. Results

In this section, we are going to present and discuss the obtained results in the calibration and verification of the proposed sensor system for the smart bottle.

A. *Descriptive analysis of calibration results*

First of all, the average values of the calibration results for each sample and the standard deviation are presented in Figure 2. In general terms, the analogRead() value of the IR light is the largest in terms of deviations and gathered values. The tea is the sample that registered the largest average value and standard deviation, 1460 ± 72.

In all cases, the higher the value, the greater the amount of light that reaches the photoreceptor. Since milk is a colloidal dilution, the amount of light that it absorbs is greater than that of water. In both water and milk infusions, the addition of the infusions decreases the amount of light that reaches the photodetectors in barely all the lights. In the case of water, this effect is more evident with green, blue, and cian lights, which might be related to the infusions' pigments.

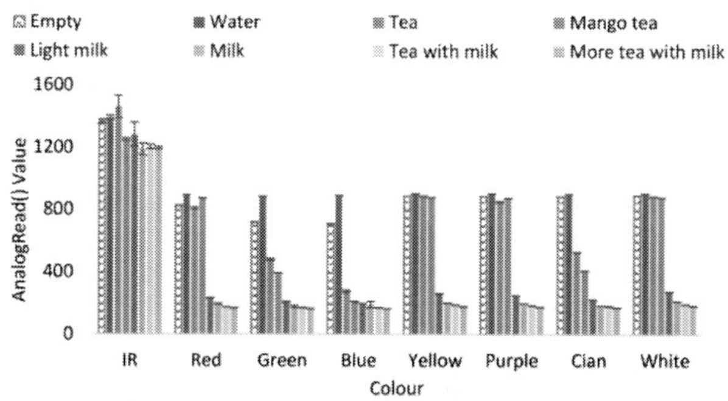

Figure 2. Average and standard deviation of calibration tests.

In order to evaluate which one of the included lights offers better results to assess the classification of drinks included in the bottle, a multivariate analysis is performed to obtain the correlation matrix. The Sample IDs have been configured according to the values obtained from the analyses of the average data. The results of the correlation matrix with the Pearson product-moment correlation coefficient, see Figure 3, indicated that the lights with the highest correlation with the Sample ID are white, purple, yellow, and green. Thus, these lights will be used for further tests.

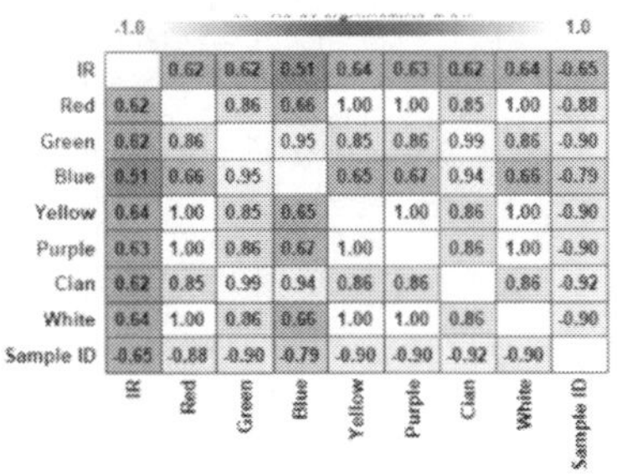

Figure 3. Correlation matrix.

B. Sample classification using thresholds based on ANOVA results

The first classification method tested is the use of a series of thresholds based on the results of the ANOVA and the multiple ranges tests. For this purpose, the results of the ANOVAs for the four selected lights are compared in Table 3. Even with all the lights, the obtained p-value is below 0.001. The multiple ranges tests offered different results for the analyzed lights. The results indicate that the white and purple lights are the ones that correctly divided the samples into individual groups. Using the yellow light and the results of the ANOVA, it is impossible to distinguish Samples IDs 1 and 3, which can provoke that the system is incapable of distinguishing an empty bottle from a bottle with tea. Meanwhile, with the results when the green light is used, the system can confuse Samples IDs 7 and 6 and 7 and 8. It will lead to the incapacity of differentiation between milk with tea and milk or milk with a stronger tea.

Thus, the white and purple lights are selected to be studied in depth in the following subsection. Even though all the lights can be used, the aim of reducing the required lights is to reduce the energy consumption of the bottle in both data gathering and data processing. Table 4 summarizes the threshold values for the classification with the ANOVA results.

To evaluate the degree of efficiency of classification using the ANOVA thresholds, two alternative and well-known methods are evaluated. The classification results with ANN and DA can be seen in Figures 4 and 5. In this case, two different classification approaches have been followed. On the one hand, all gathered data, including all light sources, is used. As in the previous case, the four calibration repetitions are used for the training dataset. On the other hand, a second classification is conducted using only the data of purple and white light according to the results of the ANOVA and multiple ranges tests.

In Figure 4, we can see the confusion matrixes of the training dataset for DA when all lights are used a) and with selected lights b). It is possible to see that there are no differences between both classifications.

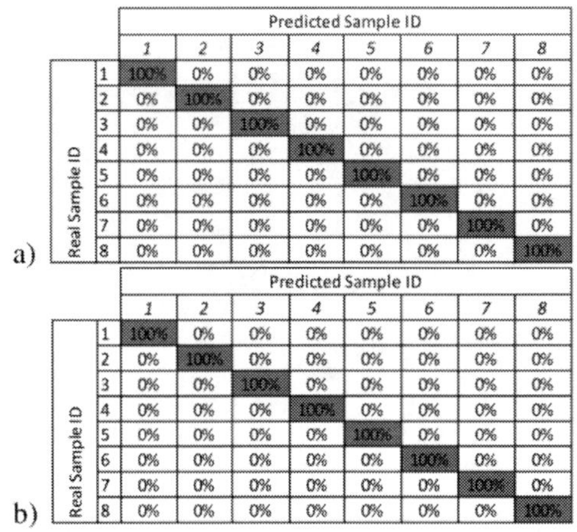

Figure 4. Confusion matrixes with DA when all data is used a) and when selected data is used b).

On the contrary, in Figure 5, the results of the obtained confusion matrixes with ANN when all data is selected a) and only white and purple data are included b) have a great variation. When all lights are used, there is a high classification error in the samples based on milk (including light milk, milk and milk with teas). This error is absent when filtered data, including only the values obtained with the purple and white light.

TABLE III. SUMMARY OF ANOVA RESULTS

Light	Sample ID								p-value
	1	*2*	*3*	*4*	*5*	*6*	*7*	*8*	
White	890.0 ^{b}	902.0 ^{a}	883.75 ^{c}	876.25 ^{d}	273.0 ^{e}	211.75 ^{f}	193.75 ^{g}	181.5 ^{h}	<0.0001
Purple	886.25 ^{b}	900.25 ^{a}	848.24 ^{c}	872 ^{d}	247.25 ^{e}	197.5 ^{f}	184.0 ^{g}	174.5 ^{h}	<0.0001
Yellow	886.25 ^{b}	900 ^{a}	885.25 ^{b}	877.25 ^{c}	260.25 ^{d}	200.25 ^{e}	186 ^{f}	177 ^{g}	<0.0001
Green	717.5 ^{b}	884.25 ^{a}	479.5 ^{c}	390.75 ^{d}	207.75 ^{e}	178.75 ^{f}	171.5 ^{fg}	164.5 ^{g}	<0.0001

a. Different letters indicate different groups.

TABLE IV. TRESHODLS FOR CLASSIFICATION WITH ANOVA RESULTS

Sample ID	White thresholds		Purple thresholds	
	Maximum	*Minimum*	*Maximum*	*Minimum*
1	1024.0	893.4	1024.0	896.1
2	893.3	879.2	896.0	887.0
3	879.1	853.1	886.9	880.1
4	867.2	536.0	880.0	574.7
5	559.6	222.5	574.6	242.5
6	222.4	190.9	242.4	202.9
7	190.8	179.4	202.8	187.7
8	179.3	0.0	187.6	0.0

C. Sample classification using DA and ANN

a)

b)

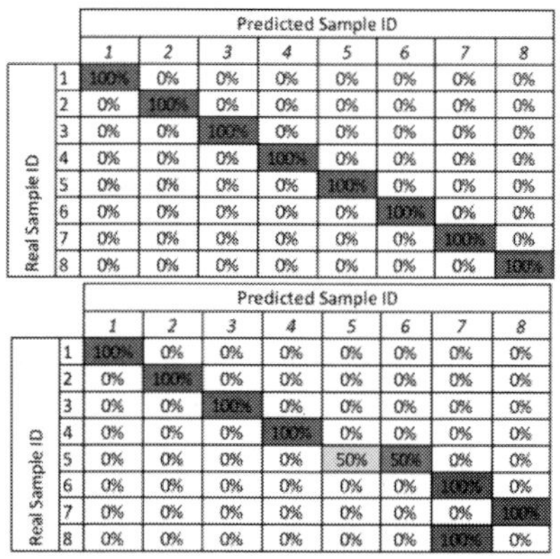

Figure 5. Confusion matrixes with DA when all data is used a) and when selected data is used b).

D. Verification

In order to decide which method should be used to determine the content of the bottle, the three presented methods in the classification are compared. Confusion matrixes are used to compare the results of the verification.

Figure 6 contains the confusion matrixes using the thresholds obtained with the ANOVA for purple a) and white b) data. It is possible to see that among the individual lights, the white one offered better results. The classification errors are linked to confusing tea with mango tea and milk with milk and tea.

Regarding the results of DA, Figure 7 presents the confusion matrixes of the verification test with all a) and selected data b). The classification is better when only selected data (purple and white combined lights) are used in DA. If all data is used, there are misclassifications between milk and different tea types.

Finally, the verification test results based on ANN with all a) and selected data b) can be seen in Figure 8. Again, the results indicate that the classification is more accurate when only white and purple light data is used. Nevertheless, in this case, the error is lesser than the errors in the training dataset classification. The error is limited to the sample composed of milk, Sample ID equal to 6, which is confused with milt and tea.

a)

b)

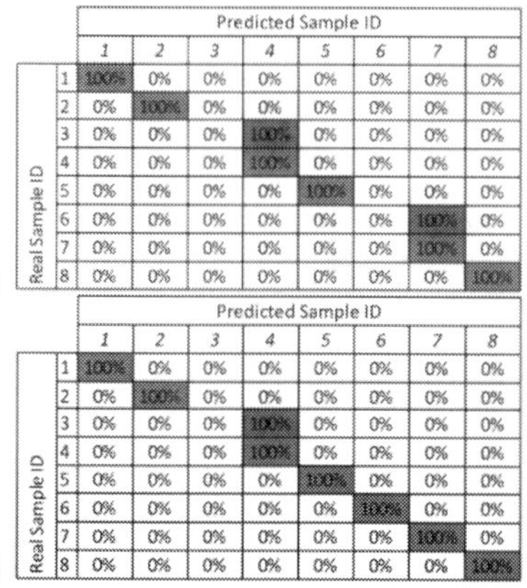

Figure 6. Confusion matrixes of verification test with purple a) and white b) individual data based on ANOVA thresholds.

a)

b)

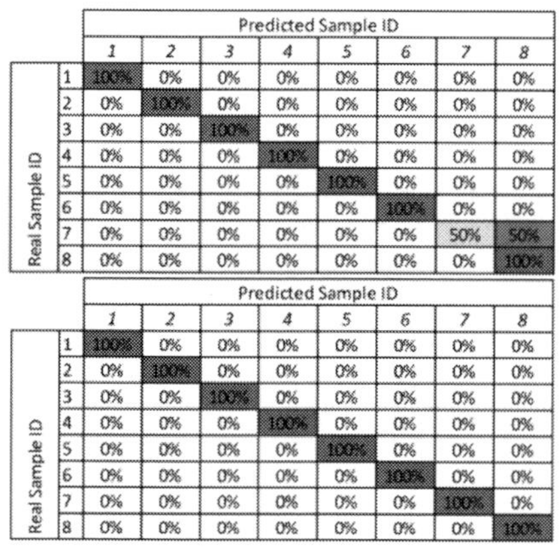

Figure 7. Confusion matrixes of verification test with all a) and selected b) data based DA.

a)

b)

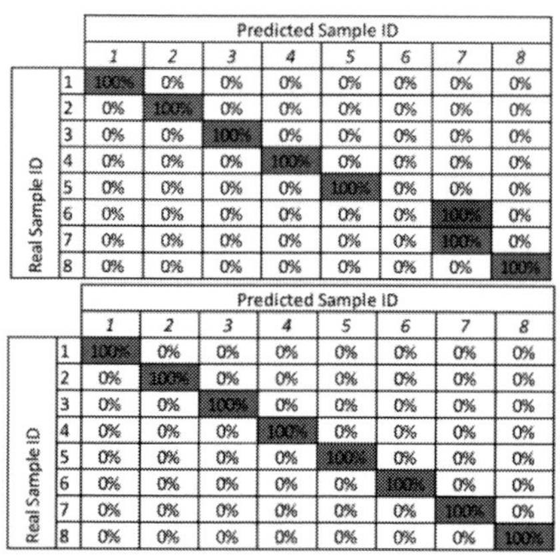

Figure 8. Confusion matrixes of verification test with all a) and selected b) data based ANN.

Thus, we can conclude that the best classification method is the DA. Although the best results are obtained with DA, ANOVA thresholds can be used if there is no need for differentiation between used teas.

VI. CONCLUSIONS

The research seeks to enhance current liquid intake detection systems by introducing an RGB sensor for discriminating liquids, facilitating precise hydration assessment. Different approaches for data classification are compared.

Results demonstrate successful differentiation using a limited number of light wavelengths, primarily white and purple. Discriminant Analysis stands out as the optimal classification method, with 100% of cases correctly classified in the verification phase. Although ANOVA thresholds can be used to achieve similar results with less computational demand when differentiating teas isn't essential.

Future work involves including additional sensors in the smart bottle to detect abnormalities in the liquids. Moreover, the integration of the smart bottle with other devices, creating an IoT solution for ambient assisted living and eHealth for elderly people, is foreseen as part of the ongoing projects.

ACKNOWLEDGMENT

This work is partially funded by the "Programa Estatal de I + D + i Orientada a los Retos de la Sociedad, en el marco del Plan Estatal de Investigación Científica y Técnica y de Innovación 2017-2020 project PID2020-114467RR-C33/AEI/10.13039/501100011033, by "Proyectos Estratégicos Orientados a la Transición Ecológica y a la Transición Digital" project TED2021-131040B-C31, and by the ThinkInAzul programme and was supported by MCIN with funding from European Union NextGenerationEU (PRTR-C17.I1) and by Generalitat Valenciana (THINKINAZUL/2021/002).

REFERENCES

[1] R. Cohen, G. Fernie, and A. Roshan Fekr, "Monitoring fluid intake by commercially available smart water bottles," Scientific Reports, vol. 12, no. 1, pp. 4402, 2022.

[2] M. Pedram, et al. "Lids: Mobile system to monitor type and volume of liquid intake," IEEE Sensors Journal, vol. 21, no. 18, pp. 20750-20763, 2021.

[3] N. Pushpalatha et al. "A Review on Intelligent Water Bottle Powered by IoT,". In Proceedings of the 2023 9th International Conference on Advanced Computing and Communication Systems (ICACCS), Coimbatore, India,17-18 March 2023, pp. 2335-2338.

[4] S. Flutura et al. "Drinkwatch: A mobile well-being application based on interactive and cooperative machine learning". In Proceedings of the 2018 International Conference on Digital Health (DH'18), Lyon, France, 23-26 April 2018, pp. 65-74.

[5] M. Rosenberg, et al. "Wearable alcohol monitors for alcohol use data collection among college students: feasibility and acceptability," Alcohol, vol. 111, pp. 75-83, 2023.

[6] M. S. Borofsky, C. A. Dauw, N. York, C. Terry, and J. E. Lingeman, "Accuracy of daily fluid intake measurements using a "smart" water bottle," Urolithiasis, vol. 46, pp. 343-348, 2018.

[7] L. B. Baker, et al. "Accuracy of a smart bottle in measuring fluid intake by American football players during pre-season training," Scientific Reports, vol. 13, no. 1, pp. 11383, 2023.

[8] R. Cohen, G. Fernie, and A. Roshan Fekr, "Fluid intake monitoring systems for the elderly: A review of the literature," Nutrients, vol. 13, no. 6, pp. 2092, 2021

[9] E. Jovanov, V. R. Nallathimmareddygari, and J. E. Pryor, "SmartStuff: A case study of a smart water bottle," In 38th Annual International Conference of the IEEE Engineering in Medicine and Biology Society (EMBC'16), Orlando, USA, 16-20 August 2016, pp. 6307-6310

[10] T. E. Stout, et al "A randomized trial Evaluating the Use of a smart water bottle to increase fluid intake in stone formers," Journal of Renal Nutrition, vol. 32, no. 4, pp. 389-395, 2022.

[11] S. Sendra, L. Parra, V. Ortuño, and J. Lloret, "A low cost turbidity sensor development," in Proceedings of the Seventh International Conference on Sensor Technologies and Applications (SENSORCOMM 2013), Barcelona, Spain, 25-31 August 2013, pp. 25–31

[12] L. Parra, L., S. Viciano-Tudela, D. Carrasco, S. Sendra, and J. Lloret, "Low-cost microcontroller-based multiparametric probe for coastal area monitoring," Sensors, 23(4), p. 1871, 2023.